Stories of Inv

The Adventures of Invento

Doubleday

Alpha Editions

This edition published in 2024

ISBN : 9789362925671

Design and Setting By
Alpha Editions
www.alphaedis.com
Email - info@alphaedis.com

Contents

INTRODUCTION

There are many thrilling incidents—all the more attractive because of their truth—in the study, the trials, the disappointments, the obstacles overcome, and the final triumph of the successful inventor.

Every great invention, afterward marvelled at, was first derided. Each great inventor, after solving problems in mechanics or chemistry, had to face the jeers of the incredulous.

The story of James Watt's sensations when the driving-wheels of his first rude engine began to revolve will never be told; the visions of Robert Fulton, when he puffed up the Hudson, of the fleets of vessels that would follow the faint track of his little vessel, can never be put in print.

It is the purpose of this book to give, in a measure, the adventurous side of invention. The trials and dangers of the builders of the submarine; the triumphant thrill of the inventor who hears for the first time the vibration of the long-distance message through the air; the daring and tension of the engineer who drives a locomotive at one hundred miles an hour.

The wonder of the mechanic is lost in the marvel of the machine; the doer is overshadowed by the greatness of his achievement.

These are true stories of adventure in invention.

HOW GUGLIELMO MARCONI TELEGRAPHS WITHOUT WIRES

A nineteen-year-old boy, just a quiet, unobtrusive young fellow, who talked little but thought much, saw in the discovery of an older scientist the means of producing a revolutionising invention by which nations could talk to nations without the use of wires or tangible connection, no matter how far apart they might be or by what they might be separated. The possibilities of Guglielmo (William) Marconi's invention are just beginning to be realised, and what it has already accomplished would seem too wonderful to be true if the people of these marvellous times were not almost surfeited with wonders.

It is of the boy and man Marconi that this chapter will tell, and through him the story of his invention, for the personality, the talents, and the character of the inventor made wireless telegraphy possible.

It was an article in an electrical journal describing the properties of the "Hertzian waves" that suggested to young Marconi the possibility of sending messages from one place to another without wires. Many men doubtless read the same article, but all except the young Italian lacked the training, the power of thought, and the imagination, first to foresee the great things that could be accomplished through this discovery, and then to study out the mechanical problem, and finally to steadfastly push the work through to practical usefulness.

It would seem that Marconi was not the kind of boy to produce a revolutionising invention, for he was not in the least spectacular, but, on the contrary, almost shy, and lacking in the aggressive enthusiasm that is supposed to mark the successful inventor; quiet determination was a strong characteristic of the young Italian, and a studious habit which had much to do with the great results accomplished by him at so early an age.

He was well equipped to grapple with the mighty problem which he had been the first to conceive, since from early boyhood he had made electricity his chief study, and a comfortable income saved him from the grinding struggle for bare existence that many inventors have had to endure. Although born in Bologna (in 1874) and bearing an Italian name, Marconi is half Irish, his mother being a native of Britain. Having been educated in Bologna, Florence, and Leghorn, Italy's schools may rightly claim to have had great influence in the shaping of his career. Certain it is, in any case, that he was well educated, especially in his chosen branch.

Marconi, like many other inventors, did not discover the means by which the end was accomplished; he used the discovery of other men, and turned their impractical theories and inventions to practical uses, and, in addition, invented many theories of his own. The man who does old things in a new way, or makes new uses of old inventions, is the one who achieves great things. And so it was the reading of the discovery of Hertz that started the boy on the train of thought and the series of experiments that ended with practical, everyday telegraphy without the use of wires. To begin with, it is necessary to give some idea of the medium that carries the wireless messages.It is known that all matter, even the most compact and solid of substances, is permeated by what is called ether, and that the vibrations that make light, heat, and colour are carried by this mysterious substance as water carries the wave motions on its surface. This strange substance, ether, which pervades everything, surrounds everything, and penetrates all things, is mysterious, since it cannot be seen nor felt, nor made known to the human senses in any way; colourless, odourless, and intangible in every way, its properties are only known through the things that it accomplishes that are beyond the powers of the known elements. Ether has been compared by one writer to jelly which, filling all space, serves as a setting for the planets, moons, and stars, and, in fact, all solid substances; and as a bowl of jelly carries a plum, so all solid things float in it.

Heinrich Hertz discovered that in addition to the light, heat, and colour waves carried by ether, this substance also served to carry electric waves or vibrations, so that electric impulses could be sent from one place to another without the aid of wires. These electric waves have been named "Hertzian waves," in honour of their discoverer; but it remained for Marconi, who first conceived their value, to put them to practical use. But for a year he did not attempt to work out his plan, thinking that all the world of scientists were studying the problem. The expected did not happen, however. No news of wireless telegraphy reached the young Italian, and so he set to work at his father's farm in Bologna to develop his idea.

THE MARCONI STATION AT GLACE BAY, CAPE BRETON

And so the boy began to work out his great idea with a dogged determination to succeed, and with the thought constantly in mind spurring him on that it was more than likely that some other scientist was striving with might and main to gain the same end.

His father's farm was his first field of operations, the small beginnings of experiments that were later to stretch across many hundreds of miles of ocean. Set up on a pole planted at one side of the garden, he rigged a tin box to which he connected, by an insulated wire, his rude transmitting apparatus. At the other side of the garden a corresponding pole with another tin box was set up and connected with the receiving apparatus. The interest of the young inventor can easily be imagined as he sat and watched for the tick of his recording instrument that he knew should come from the flash sent across the garden by his companion. Much time had been spent in the planning and the making of both sets of instruments, and this was the first test; silent he waited, his nerves tense, impatient, eager. Suddenly the Morse sounder began to tick and burr-r-r; the boy's eyes flashed, and his heart gave an exultant bound—the first wireless message had been sent and received, and a new marvel had been added to the list of world's wonders. The quiet farm was the scene of many succeeding experiments, the place having been put at his disposal by his appreciative father, and in addition ample funds were generously supplied from the same source. Different heights of poles were tried, and it was found that the distance could be increased in proportion to the altitude of the pole bearing the receiving and transmitting tin boxes or "capacities"—the higher the poles the greater distance the message could be sent. The success of Marconi's system depended largely on his receiving apparatus, and it is on account of his use of some of the devices invented by other men that unthinking people have criticised him. He adapted to the use of wireless telegraphy certain inventions that had heretofore been merely interesting scientific toys—curious little instruments of no apparent practical value until his eye saw in them a contributory means to a great end.

Though Hertz caught the etheric waves on a wire hoop and saw the answering sparks jump across the unjoined ends, there was no way to record the flashes and so read the message. The electric current of a wireless message was too weak to work a recording device, so Marconi made use of an ingenious little instrument invented by M. Branly, called a coherer, to hitch on, as it were, the stronger current of a local battery. So the weak current of the ether waves, aided by the stronger current of the local circuit, worked the recorder and wrote the message down. The coherer was a little tube of glass not as long as your finger, and smaller than a lead pencil, into each end of which was tightly fitted plugs of silver; the plugs met within a small fraction of an inch in the centre of the tube, and the very small space between the ends of the plugs was filled with silver and nickel dust so fine as to be almost as light as air. Though a small instrument, and more delicate than a clinical thermometer, it loomed large in the working-out of wireless telegraphy. One of the silver plugs of the coherer was connected to the receiving wire, while the other was connected to the earth (grounded). To one plug of the coherer also was joined one pole of the local battery, while the other pole was in circuit with the other plug of the coherer through the recording instrument. The fine dust-like silver and nickel particles in the coherer possessed the quality of high resistance, except when charged by the electric current of the ether waves; then the particles of metal clung together, cohered, and allowed of the passage of the ether waves' current and the strong current of the local battery, which in turn actuated the Morse sounder and recorder. The difficulty with this instrument was in the fact that the metal particles continued to cohere, unless shaken apart, after the ether waves' current was discontinued. So Marconi invented a little device which was in circuit with the recorder and tapped the coherer tube with a tiny mallet at just the right moment, causing the particles to separate, or decohere, and so break the circuit and stop the local battery current. As no wireless message could have been received without the coherer, so no record or reading could have been made without the young Italian's improvement.

In sending the message from one side of his father's estate at Bologna to the other the young inventor used practically the same methods that he uses to-day. Marconi's transmitting apparatus consisted of electric batteries, an induction coil by which the force of the current is increased, a telegrapher's key to make and break the circuit, and a pair of brass knobs. The batteries were connected with the induction coil, which in turn was connected with the brass knobs; the telegrapher's key was placed between the battery and the coil. It was the boy scarcely out of his teens who worked out the principles of his system, but it took time and many, many experiments to overcome the obstacles of long-distance wireless telegraphy. The sending of a message across the garden in far-away Italy was a simple matter—the depressed key completed the electric circuit created by a strong battery through the induction coil and made a spark jump between the two brass knobs, which in turn started the ether vibrating at the rate of three or four hundred million times a minute from the tin box on top of a pole. The vibrations in the ether circled wider and wider, as the circular waves spread from the spot where a stone is dropped into a pool, but with the speed of light, until they reached a corresponding tin box on top of a like pole on the other side of the garden; this box, and the wire connected with it, caught the waves, carried them down to the coherer, and, joining the current from the local battery, a dot or dash was recorded; immediately after, the tapper separated the metal particles in the coherer and it was ready for the next series of waves.

One spark made a single dot, a stream of sparks the dash of the Morse telegraphic code. The apparatus was crude at first, and worked spasmodically, but Marconi knew he was on the right track and persevered. With the heightening of the pole he found he could send farther without an increase of electric power, until wireless messages were sent from one extreme limit of his father's farm to the other.

It is hard to realize that the young inventor only began his experiments in wireless telegraphy in 1895, and that it is scarcely eight years since the great idea first occurred to him.

After a year of experimenting on his father's property, Marconi was able to report to W.H. Preece, chief electrician of the British postal system, certain definite facts—not theories, but facts. He had actually sent and received messages, without the aid of wires, about two miles, but the facilities for further experimenting at Bologna were exhausted, and he went to England.

Here was a youth (scarcely twenty-one), with a great invention already within his grasp—a revolutionising invention, the possibilities of which can hardly yet be conceived. And so this young Italian, quiet, retiring, unassuming, and yet possessing Jove's power of sending thunderbolts, came to London (in 1896), to upbuild and link nation to nation more closely. With his successful experiments behind him, Marconi was well received in England, and began his further work with all the encouragement possible. Then followed a series of tests that were fairly bewildering. Messages were sent through brick walls—through houses, indeed—over long stretches of plain, and even through hills, proving beyond a doubt that the etheric electric waves penetrated everything. For a long time Marconi used modifications of the tin boxes which were a feature of his early trials, but later balloons covered with tin-foil, and then a kite six feet high, covered with thin metallic sheets, was used, the wire leading down to the sending and receiving instruments running down the cord. With the kite, signals were sent eight miles by the middle of 1897. Marconi was working on the theory that the higher the transmitting and receiving "capacity," as it was then called, or wire, or "antenna," the greater distance the message could be sent; so that the distance covered was only limited by the height of the transmitting and receiving conductors. This theory has since been abandoned, great power having been substituted for great height.

Marconi saw that balloons and kites, the playthings of the winds, were unsuitable for his purpose, and sought some more stable support for his sending and receiving apparatus. He set up, therefore (in November, 1897), at the Needles, Isle of Wight, a 120-foot mast, from the apex of which was strung his transmitting wire (an insulated wire, instead of a box, or large metal body, as heretofore used). This was the forerunner of all the tall spars that have since pointed to the sky, and which have been the centre of innumerable etheric waves bearing man's messages over land and sea.

With the planting of the mast at the Needles began a new series of experiments which must have tried the endurance and determination of the young man to the utmost. A tug was chartered, and to the sixty-foot mast erected thereon was connected the wire and transmitting and receiving apparatus. From this little vessel Marconi sent and received wireless signals day after day, no matter what the state of the weather. With each trip experience was accumulated and the apparatus was improved; the moving station steamed farther and farther out to sea, and the ether waves circled wider and wider, until, at the end of two months of sea-going, wireless telegraphy signals were received clear across to the mainland, fourteen miles, whereupon a mast was set up and a station established (at Bournemouth), and later eighteen miles away at Poole.

By the middle of 1898 Marconi's wireless system was doing actual commercial service in reporting, for a Dublin newspaper, the events at a regatta at Kingstown, when about seven hundred messages were sent from a floating station to land, at a maximum distance of twenty-five miles.

It was shortly afterward, while the royal yacht was in Cowes Bay, that one hundred and fifty messages between the then Prince of Wales and his royal mother at Osborne House were exchanged, most of them of a very private nature.

One of the great objections to wireless telegraphy has been the inability to make it secret, since the ether waves circle from the centre in all directions, and any receiving apparatus within certain limits would be affected by the waves just as the station to which the message was sent would be affected by them. To illustrate: the waves radiating from a stone dropped into a still pool would make a dead leaf bob up and down anywhere on the pool within the circle of the waves, and so the ether waves excited the receiving apparatus of any station within the effective reach of the circle.

Of course, the use of a cipher code would secure the secrecy of a message, but Marconi was looking for a mechanical device that would make it impossible for any but the station to which the message was sent to receive it. He finally hit upon the plan of focussing the ether waves as the rays of a searchlight are concentrated in a given direction by the use of a reflector, and though this adaptation of the searchlight principle was to a certain extent successful, much penetrating power was lost. This plan has been abandoned for one much more ingenious and effective, based on the principle of attunement, of which more later.

It was a proud day for the young Italian when his receiver at Dover recorded the first wireless message sent across the British Channel from Boulogne in 1899—just the letters V M and three or four words in the Morse alphabet of dots and dashes. He had bridged that space of stormy, restless water with an invisible, intangible something that could be neither seen, felt, nor heard, and yet was stronger and surer than steel—a connection that nothing could interrupt, that no barrier could prevent. The first message from England to France was soon followed by one to M. Branly, the inventor of the coherer, that made the receiving of the message possible, and one to the queen of Marconi's country. The inventor's march of progress was rapid after this—stations were established at various points all around the coast of England; vessels were equipped with the apparatus so that they might talk to the mainland and to one another. England's great dogs of war, her battle-ships, fought an imaginary war with one another and the orders were flashed from the flagship to the fighters, and from the Admiral's cabin to the shore, in spite of fog and great stretches of open water heaving between.

THE WIRELESS TELEGRAPH STATION AT GLACE BAY

A lightship anchored off the coast of England was fitted with the Marconi apparatus and served to warn several vessels of impending danger, and at last, after a collision in the dark and fog, saved the men who were aboard of her by sending a wireless message to the mainland for help.

From the very beginning Marconi had set a high standard for himself. He worked for an end that should be both commercially practical and universal. When he had spanned the Channel with his wireless messages, he immediately set to work to fling the ether waves farther and farther. Even then the project of spanning the Atlantic was in his mind.

On the coast of Cornwall, near Penzance, England, Marconi erected a great station. A forest of tall poles were set up, and from the wires strung from one to the other hung a whole group of wires which were in turn connected to the transmitting apparatus. From a little distance the station looked for all the world like ships' masts that had been taken out and ranged in a circle round the low buildings. This was the station of Poldhu, from which Marconi planned to send vibrations in the ether that would reach clear across to St. Johns, Newfoundland, on the other side of the Atlantic—more than two thousand miles away. A power-driven dynamo took the place of the more feeble batteries at Poldhu, converters to increase the power displaced the induction coil, and many sending-wires, or antennae, were used instead of one.

On Signal Hill, at St. Johns, Newfoundland—a bold bluff overlooking the sea—a group of men worked for several days, first in the little stone house at the brink of the bluff, setting up some electric apparatus; and later, on the flat ground nearby, the same men were very busy flying a great kite and raising a balloon. There was no doubt about the earnestness of these men: they were not raising that kite for fun. They worked with care and yet with an eagerness that no boy ever displays when setting his home-made or store flyer to the breeze. They had hard luck: time and time again the wind or the rain, or else the fog, baffled them, but a quiet young fellow with a determined, thoughtful face urged them on, tugged at the cord, or held the kite while the others ran with the line. Whether Marconi stood to one side and directed or took hold with his men, there was no doubt who was master. At last the kite was flying gallantly, high overhead in the blue. From the sagging kite-string hung a wire that ran into the low stone house.

One cold December day in 1901, Guglielmo Marconi sat still in a room in the Government building at Signal Hill, St. Johns, Newfoundland, with a telephone receiver at his ear and his eye on the clock that ticked loudly nearby. Overhead flew his kite bearing his receiving-wire. It was 12:30 o'clock on the American side of the ocean, and Marconi had ordered his operator in far-off Poldhu, two thousand watery miles away, to begin signalling the letter "S"—three dots of the Morse code, three flashes of the bluish sparks—at that corresponding hour. For six years he had been looking forward to and working for that moment—the final test of all his effort and the beginning of a new triumph. He sat waiting to hear three small sounds, the br-br-br of the Morse code "S," humming on the diaphragm of his receiver—the signature of the ether waves that had travelled two thousand miles to his listening ear. As the hands of the clock, whose ticking alone broke the stillness of the room, reached thirty minutes past twelve, the receiver at the inventor's ear began to hum, br-br-br, as distinctly as the sharp rap of a pencil on a table—the unmistakable note of the ether vibrations sounded in the telephone receiver. The telephone receiver was used instead of the usual recorder on account of its superior sensitiveness.

Transatlantic wireless telegraphy was an accomplished fact.

Though many doubted that an actual signal had been sent across the Atlantic, the scientists of both continents, almost without exception, accepted Marconi's statement. The sending of the transatlantic signal, the spanning of the wide ocean with translatable vibrations, was a great achievement, but the young Italian bore his honours modestly, and immediately went to work to perfect his system.

Two months after receiving the message from Poldhu at St. Johns, Marconi set sail from England for America, in the *Philadelphia*, to carry out, on a much larger scale, the experiments he had worked out with the tug three years ago. The steamship was fitted with a complete receiving and sending outfit, and soon after she steamed out from the harbor she began to talk to the Cornwall station in the dot-and-dash sign language. The long-distance talk between ship and shore continued at intervals, the recording instrument writing the messages down so that any one who understood the Morse code could read. Message after message came and went until one hundred and fifty miles of sea lay between Marconi and his station. Then the ship could talk no more, her sending apparatus not being strong enough; but the faithful men at Poldhu kept sending messages to their chief, and the recorder on the *Philadelphia* kept taking them down in the telegrapher's shorthand, though the steamship was plowing westward at twenty miles an hour. Day after day, at the appointed hour to the very second, the messages came from the station on land, flung into the air with the speed of light, to the young man in the deck cabin of a speeding steamship two hundred and fifty, five hundred, a thousand, fifteen hundred, yes, two thousand and ninety-nine miles away—messages that were written down automatically as they came, being permanent records that none might gainsay and that all might observe.

To Marconi it was the simple carrying out of his orders, for he said that he had fitted the Poldhu instruments to work to two thousand one hundred miles, but to those who saw the thing done—saw the narrow strips of paper come reeling off the recorder, stamped with the blue impressions of the messages through the air, it was astounding almost beyond belief; but there was the record, duly attested by those who knew, and clearly marked with the position of the ship in longitude and latitude at the time they were received.

It was only a few months afterward that Marconi, from his first station in the United States, at Wellfleet, Cape Cod, Mass., sent a message direct to Poldhu, three thousand miles. At frequent intervals messages go from one country to the other across the ocean, carried through fog, unaffected by the winds, and following the curvature of the earth, without the aid of wires.

Again the unassuming nature of the young Italian was shown. There was no brass band nor display of national colours in honour of the great achievement; it was all accomplished quietly, and suddenly the world woke up to find that the thing had been done. Then the great personages on both sides of the water congratulated and complimented each other by Marconi's wireless system.

At Marconi's new station at Glacé Bay, Cape Breton, and at the powerful station at Wellfleet, Cape Cod, the receiving and sending wires are supported by four great towers more than two hundred feet high. Many wires are used instead of one, and much greater power is of course employed than at first, but the marvellously simple principle is the same that was used in the garden at Bologna. The coherer has been displaced by a new device invented by Marconi, called a magnetic detector, by which the ether waves are aided by a stronger current to record the message. The effect is the same, but the method is entirely different.

The sending of a long-distance message is a spectacular thing. Current of great power is used, and the spark is a blinding flash accompanied by deafening noises that suggest a volley from rifles. But Marconi is experimenting to reduce the noise, and the use of the mercury vapour invented by Peter Cooper Hewitt will do much to increase the rapidity in sending.

After much experimenting Marconi discovered that the longer the waves in the ether the more penetrating and lasting the quality they possessed, just as long swells on a body of water carry farther and endure longer than short ones. Moreover, he discovered that if many sending-wires were used instead of one, and strong electric power was employed instead of weak, these long, penetrating, enduring waves could be produced. All the new Marconi stations, therefore, built for long-distance work, are fitted with many sending-wires, and powerful dynamos are run which are capable of producing a spark between the silvered knobs as thick as a man's wrist.

Marconi and several other workers in the field of wireless telegraphy are now busy experimenting on a system of attunement, or syntony, by which it will be possible to so adjust the sending instruments that none but the receiver for whom the message is meant can receive it. He is working on the principle whereby one tuning-fork, when set vibrating, will set another of the same pitch humming. This problem is practically solved now, and in the near future every station, every ship, and each installation will have its own key, and will respond to none other than the particular vibrations, wave lengths, or oscillations, for which it is adjusted.

All through the wonders he has brought about, Marconi, the boy and the man, has shown but little—he is the strong character that does things and says little, and his works speak so amazingly, so loudly, that the personality of the man is obscured.

The Marconi station at Glacé Bay, Cape Breton, is now receiving messages for cableless transmission to England at the rate of ten cents a word—newspaper matter at five cents a word. Transatlantic wireless telegraphy is an everyday occurrence, and the common practical uses are almost beyond mention. It is quite within the bounds of possibility for England to talk clear across to Australia over the Isthmus of Panama, and soon France will be actually holding converse with her strange ally, Russia, across Germany and Austria, without asking the permission of either country. Ships talk to one another while in mid-ocean, separated by miles of salt water. Newspapers have been published aboard transatlantic steamers with the latest news telegraphed while en route; indeed, a regular news service of this kind, at a very reasonable rate, has been established. These are facts; what wonders the future has in store we can only guess. But these are some of the possibilities—news service supplied to subscribers at their homes, the important items to be ticked off on each private instrument automatically, "Marconigraphed" from the editorial rooms; the sending and receiving of messages from moving trains or any other kind of a conveyance; the direction of a submarine craft from a safe-distance point, or the control of a submarine torpedo.

One is apt to grow dizzy if the imagination is allowed to run on too far—but why should not one friend talk to another though he be miles away, and to him alone, since his portable instrument is attuned to but one kind of vibration. It will be like having a separate language for each person, so that "friend communeth with friend, and a stranger intermeddleth not—" and which none but that one person can understand.

SANTOS-DUMONT AND HIS AIR-SHIP

There was a boy in far-away Brazil who played with his friends the game of "Pigeon Flies."

In this pastime the boy who is "it" calls out "pigeon flies," or "bat flies," and the others raise their fingers; but if he should call "fox flies," and one of his mates should raise his hand, that boy would have to pay a forfeit.

The Brazilian boy, however, insisted on raising his finger when the catchwords "man flies" were called, and firmly protested against paying a forfeit.

Alberto Santos-Dumont, even in those early days, was sure that if man did not fly then he would some day.

Many an imaginative boy with a mechanical turn of mind has dreamed and planned wonderful machines that would carry him triumphantly over the tree-tops, and when the tug of the kite-string has been felt has wished that it would pull him up in the air and carry him soaring among the clouds. Santos-Dumont was just such a boy, and he spent much time in setting miniature balloons afloat, and in launching tiny air-ships actuated by twisted rubber bands. But he never outgrew this interest in overhead sailing, and his dreams turned into practical working inventions that enabled him to do what never a mortal man had done before—that is, move about at will in the air.

Perhaps it was the clear blue sky of his native land, and the dense, almost impenetrable thickets below, as Santos-Dumont himself has suggested, that made him think how fine it would be to float in the air above the tangle, where neither rough ground nor wide streams could hinder. At any rate, the thought came into the boy's mind when he was very small, and it stuck there.

His father owned great plantations and many miles of railroad in Brazil, and the boy grew up in the atmosphere of ponderous machinery and puffing locomotives. By the time Santos-Dumont was ten years old he had learned enough about mechanics to control the engines of his father's railroads and handle the machinery in the factories. The boy had a natural bent for mechanics and mathematics, and possessed a cool courage that made him appear almost phlegmatic. Besides his inherited aptitude for mechanics, his father, who was an engineer of the Central School of Arts and Manufactures of Paris, gave him much useful instruction. Like Marconi, Santos-Dumont had many advantages, and also, like the inventor of wireless telegraphy, he had the high intelligence and determination to win success in spite of many discouragements. Like an explorer in a strange land, Santos-Dumont was a pioneer in his work, each trial being different from any other, though the means in themselves were familiar enough.

SANTOS-DUMONT PREPARING FOR A FLIGHT IN "SANTOS-DUMONT NO. [illegible]"
[illegible]

The boy Santos-Dumont dreamed air-ships, planned air-ships, and read about aerial navigation, until he was possessed with the idea that he must build an air-ship for himself.

He set his face toward France, the land of aerial navigation and the country where light motors had been most highly developed for automobiles. The same year, 1897, when he was twenty-four years old, he, with M. Machuron, made his first ascent in a spherical balloon, the only kind in existence at that time. He has described that first ascension with an enthusiasm that proclaims him a devotee of the science for all time.

His first ascension was full of incident: a storm was encountered; the clouds spread themselves between them and the map-like earth, so that nothing could be seen except the white, billowy masses of vapour shining in the sun; some difficulty was experienced in getting down, for the air currents were blowing upward and carried the balloon with them; the tree-tops finally caught them, but they escaped by throwing out ballast, and finally landed in an open place, and watched the dying balloon as it convulsively gasped out its last breath of escaping gas.

After a few trips with an experienced aeronaut, Santos-Dumont determined to go alone into the regions above the clouds. This was the first of a series of ascensions in his own balloon. It was made of very light silk, which he could pack in a valise and carry easily back to Paris from his landing point. In all kinds of weather this determined sky navigator went aloft; in wind, rain, and sunshine he studied the atmospheric conditions, air currents, and the action of his balloon.

The young Brazilian ascended thirty times in spherical balloons before he attempted any work on an elongated shape. He realised that many things must be learned before he could handle successfully the much more delicate and sensitive elongated gas-bag.

In general, Santos-Dumont worked on the theory of the dirigible balloon—that is, one that might be controlled and made to go in any direction desired, by means of a motor and propeller carried by a buoyant gas-bag. His plan was to build a balloon, cigar-shaped, of sufficient capacity to a little more than lift his machinery and himself, this extra lifting power to be balanced by ballast, so that the balloon and the weight it carried would practically equal the weight of air it displaced. The push of the revolving propeller would be depended upon to move the whole air-ship up or down or forward, just as the motion of a fish's fins and tail move it up, down, forward, or back, its weight being nearly the same as the water it displaces.

The theory seems so simple that it strikes one as strange that the problem of aerial navigation was not solved long ago. The story of Santos-Dumont's experiments, however, his adventures and his successes, will show that the problem was not so simple as it seemed.

Santos-Dumont was built to jockey a Pegasus or guide an air-ship, for he weighed but a hundred pounds when he made his first ascensions, and added very little live ballast as he grew older.

Weight, of course, was the great bugbear of every air-ship inventor, and the chief problem was to provide a motor light enough to furnish sufficient power for driving a balloon that had sufficient lifting capacity to support it and the aeronaut in the air. Steam-engines had been tried, but found too heavy for the power generated; electric motors had been tested, and proved entirely out of the question for the same reason.

Santos-Dumont has been very fortunate in this respect, his success, indeed, being largely due to the compact and powerful gasoline motors that have been developed for use on automobiles.

Even before the balloon for the first air-ship was ordered the young Brazilian experimented with his three-and-one-half horse-power gasoline motor in every possible way, adding to its power, and reducing its weight until he had cut it down to sixty-six pounds, or a little less than twenty pounds to a horse-power. Putting the little motor on a tricycle, he led the procession of powerful automobiles in the Paris-Amsterdam race for some distance, proving its power and speed. The motor tested to his satisfaction, Santos-Dumont ordered his balloon of the famous maker, Lachambre, and while it was building he experimented still further with his little engine. To the horizontal shaft of his motor he attached a propeller made of silk stretched tightly over a light wooden framework. The motor was secured to the aeronaut's basket behind, and the reservoir of gasoline hung to the basket in front. All this was done and tested before the balloon was finished—in fact, the aeronaut hung himself up in his basket from the roof of his workshop and started his motor to find out how much pushing power it exerted and if everything worked satisfactorily.

On September 18, 1898, Santos-Dumont made his first ascension in his first air-ship—in fact, he had never tried to operate an elongated balloon before, and so much of this first experience was absolutely new. Imagine a great bag of yellow oiled silk, cigar-shaped, fully inflated with hydrogen gas, but swaying in the morning breeze, and tugging at its restraining ropes: a vast bubble eighty-two feet long, and twelve feel in diameter at its greatest girth. Such was the balloon of Santos-Dumont's first air-ship. Suspended by cords from the great gas-bag was the basket, to which was attached the motor and six-foot propeller, hung sixteen feet below the belly of the great air-fish.

Many friends and curiosity seekers had assembled to see the aeronaut make his first foolhardy attempt, as they called it. Never before had a spark-spitting motor been hung under a great reservoir of highly inflammable hydrogen gas, and most of the group thought the daring inventor would never see another sunset. Santos-Dumont moved around his suspended air-ship, testing a cord here and a connection there, for he well knew that his life might depend on such a small thing as a length of twine or a slender rod. At one side of a small open space on the outskirts of Paris the long, yellow balloon tugged at its fastenings, while the navigator made his final round to see that all was well. A twist of a strap around the driving-wheel set the motor going, and a moment later Santos-Dumont was standing in his basket, giving the signal to release the air-ship. It rose heavily, and travelling with the fresh wind, the propellers whirling swiftly, it crashed into the trees at the other side of the enclosure. The aeronaut had, against his better judgment, gone with the wind rather than against it, so the power of the propeller was added to the force of the breeze, and the trees were encountered before the ship could rise sufficiently to clear them. The damage was repaired, and two days later, September 20, 1898, the Brazilian started again from the same enclosure, but this time against the wind. The propeller whirled merrily, the explosions of the little motor snapped sharply as the great yellow bulk and the tiny basket with its human freight, the captain of the craft, rose slowly in the air. Santos-Dumont stood quietly in his basket, his hand on the controlling cords of the great rudder on the end of the balloon; near at hand was a bag of loose sand, while small bags of ballast were packed around his feet. Steadily she rose and began to move against the wind with the slow grace of a great bird, while the little man in the basket steered right or left, up or down, as he willed. He turned his rudder for the lateral movements, and changed his shifting bags of ballast hanging fore and aft, pulling in the after bag when he wished to point her nose down, and doing likewise with the forward ballast when he wished to ascend—the propeller pushing up or down as she was pointed. For the first time a man had actual control of an air-ship that carried him. He commanded it as a captain governs his ship, and it obeyed as a vessel answers its helm.

A quarter of a mile above the heads of the pygmy crowd who watched him the little South American maneuvered his air-ship, turning circles and figure eights with and against the breeze, too busy with his rudder, his vibrating little engine, his shifting bags of ballast, and the great palpitating bag of yellow silk above him, to think of his triumph, though he could still hear faintly the shouts of his friends on earth. For a time all went well and he felt the exhilaration that no earth-travelling can ever give, as he experienced somewhat of the freedom that the birds must know when they soar through the air unfettered. As he descended to a lower, denser atmosphere he felt rather than saw that something was wrong—that there was a lack of buoyancy to his craft. The engine kept on with its rapid "phut, phut, phut" steadily, but the air-ship was sinking much more rapidly than it should. Looking up, the aeronaut saw that his long gas-bag was beginning to crease in the middle and was getting flabby, the cords from the ends of the long balloon were beginning to sag, and threatened to catch in the propeller. The earth seemed to be leaping up toward him and destruction stared him in the face. A hand air-pump was provided to fill an air balloon inside the larger one and so make up for the compression of the hydrogen gas caused by the denser, lower atmosphere. He started this pump, but it proved too small, and as the gas was compressed more and more, and the flabbiness of the balloon increased, the whole thing became unmanageable. The great ship dropped and dropped through the air, while the aeronaut, no longer in control of his ship, but controlled by it, worked at the pump and threw out ballast in a vain endeavour to escape the inevitable. He was descending directly over the greensward in the centre of the Longchamps race-course, when he caught sight of some boys flying kites in the open space. He shouted to them to take hold of his trailing guide-rope and run with it against the wind. They understood at once and as instantly obeyed. The wind had the same effect on the air-ship as it has on a kite when one runs with it, and the speed of the fall was checked. Man and air-ship landed with a thud that smashed almost everything but the man. The smart boys that had saved Santos-Dumont's life helped him pack what was left of "Santos-Dumont No. 1" into its basket, and a cab took inventor and invention back to Paris.

In spite of the narrow escape and the discouraging ending of his first flight, Santos-Dumont launched his second air-ship the following May. Number 2 was slightly larger than the first, and the fault that was dangerous in it was corrected, its inventor thought, by a ventilator connecting the inner bag with the outer air, which was designed to compensate for the contraction of the gas and keep the skin of the balloon taut. But No. 2 doubled up as had No. 1, while she was still held captive by a line; falling into a tree hurt the balloon, but the aeronaut escaped unscratched. Santos-Dumont, in spite of his quiet ways and almost effeminate speech, his diminutive body, and wealth that permitted him to enjoy every luxury, persisted in his work with rare courage and determination. The difficulties were great and the available information meager to the last degree. The young inventor had to experiment and find out for himself the obstacles to success and then invent ways to surmount them. He had need of ample wealth, for the building of air-ships was expensive business. The balloons were made of the finest, lightest Japanese silk, carefully prepared and still more vigorously tested. They were made by the most famous of the world's balloon-makers, Lachambre, and required the spending of money unstintedly. The motors cost according to their lightness rather than their weight, and all the materials, cordage, metal-work, etc., were expensive for the same reason. The cost of the hydrogen gas was very great also, at twenty cents per cubic meter (thirty-five cubic feet); and as at each ascension all the gas was usually lost, the expense of each sail in the air for gas alone amounted to from $57 for the smallest ship to $122 for the largest.

SANTOS-DUMONT IN HIS AIRSHIP "NO. 6" ROUNDING THE EIFFEL TOWER ON HIS PRIZE-WINNING TRIP

Nevertheless, in November of 1899 Santos-Dumont launched another air-ship—No. 3. This one was supported by a balloon of much greater diameter, though the length remained about the same—sixty-six feet. The capacity, however, was almost three times as great as No. 1, being 17,655 cubic feet. The balloon was so much larger that the less expensive but heavier illuminating gas could be used instead of hydrogen. When the air-ship "Santos-Dumont No. 3" collapsed and dumped its navigator into the trees, Santos-Dumont's friends took it upon themselves to stop his dangerous experimenting, but he said nothing, and straightway set to work to plan a new machine. It was characteristic of the man that to him the danger, the expense, and the discouragements counted not at all.

In the afternoon of November 13, 1899, Santos-Dumont started on his first flight in No. 3. The wind was blowing hard, and for a time the great bulk of the balloon made little headway against it; 600 feet in air it hung poised almost motionless, the winglike propeller whirling rapidly. Then slowly the great balloon began nosing its way into the wind, and the plucky little man, all alone, beyond the reach of any human voice, could not tell his joy, although the feeling of triumph was strong within him. Far below him, looking like two-legged hats, so foreshortened they were from the aeronaut's point of view, were the people of Paris, while in front loomed the tall steel spire of the Eiffel Tower. To sail round that tower even as the birds float about had been the dream of the young aeronaut since his first ascension. The motor was running smoothly, the balloon skin was taut, and everything was working well; pulling the rudder slightly, Santos-Dumont headed directly for the great steel shaft.

The people who were on the Eiffel Tower that breezy afternoon saw a sight that never a man saw before. Out of the haze a yellow shape loomed larger each minute until its outlines could be distinctly seen. It was a big cigar-shaped balloon, and under it, swung by what seemed gossamer threads, was a basket in which was a man. The air-ship was going against the wind, and the man in the basket evidently had full control, for the amazed people on the tower saw the air-ship turn right and left as her navigator pulled the rudder-cords, and she rose and fell as her master regulated his shifting ballast. For twenty minutes Santos-Dumont maneuvered around the tower as a sailboat tacks around a buoy. While the people on that tall spire were still watching, the aeronaut turned his ship around and sailed off for the Longchamps race-course, the green oval of which could be just distinguished in the distance.

On the exact spot where, a little more than a year before, the same man almost lost his life and wrecked his first air-ship, No. 3 landed as softly and neatly as a bird.

Though he made many other successful flights, he discovered so many improvements that with the first small mishap he abandoned No. 3 and began on No. 4.

The balloon "Santos-Dumont No. 4" was long and slim, and had an inner air-bag to compensate for the contraction of the hydrogen gas. This air-ship had one feature that was entirely new; the aeronaut had arranged for himself, not a secure basket to stand in, but a frail, unprotected bicycle seat attached to an ordinary bicycle frame. The cranks were connected with the starting-gear of the motor.

Seated on his unguarded bicycle seat, and holding on to the handle-bars, to which were attached the rudder-cords, Santos-Dumont made voyages in the air with all the assurance of the sailor on the sea.

But No. 4 was soon too imperfect for the exacting Brazilian, and in April, 1901, he had finished No. 5. This air-cruiser was the longest of all (105 feet), and was fitted with a sixteen horse-power motor. Instead of the bicycle frame, he built a triangular keel of pine strips and strengthened it with tightly strung piano wires, the whole frame, though sixty feet long, weighing but 110 pounds. Hung between the rods, being suspended by piano wires as in a spider-web, was the motor, basket, and propeller-shaft.

The last-named air-ship was built, if not expressly at least with the intention of trying for the Deutsch Prize of 100,000 francs. This was a big undertaking, and many people thought it would never be accomplished; the successful aeronaut had to travel more than three miles in one direction, round the Eiffel Tower as a racing yacht rounds a stake-boat, and return to the starting point, all within thirty minutes—*i.e.*, almost seven miles in two directions in half an hour.

The new machine worked well, though at one time the aerial navigator's friends thought that they would have to pick him up in pieces and carry him home in a basket. This incident occurred during one of the first flights in No. 5. Everything was going smoothly, and the air-ship circled like a hawk, when the spectators, who were craning their necks to see, noticed that something was wrong; the motor slowed down, the propeller spun less swiftly, and the whole fabric began to sink toward the ground. While the people gazed, their hearts in their mouths, they saw Santos-Dumont scramble out of his basket and crawl out on the framework, while the balloon swayed in the air. He calmly knotted the cord that had parted and crept back to his place, as unconcernedly as if he were on solid ground.

It was in August of 1901 that he made his first official trial for the Deutsch Prize. The start was perfect, and the machine swooped toward the distant tower straight as a crow flies and almost as fast. The first half of the distance was covered in nine minutes, so twenty-one minutes remained for the balance of the journey: success seemed assured; the prize was almost within the grasp of the aeronaut. Of a sudden assured success was changed to dire peril; the automatic valves began to leak, the balloon to sag, the cords supporting the wooden keel hung low, and before Santos-Dumont could stop the motor the propeller had cut them and the whole system was threatened. The wind was drifting the air-ship toward the Eiffel Tower; the navigator had lost control; 500 feet below were the roofs of the Trocadero Hotels; he had to decide which was the least dangerous; there was but a moment to think. Santos-Dumont, death staring him in the face, chose the roofs. A swift jerk of a cord, and a big slit was made in the balloon. Instantly man, motor, gas-bag, and keel went tumbling down straight into the court of the hotels. The great balloon burst with a noise like an explosion, and the man was lost in a confusion of yellow-silk covering, cords, and wires. When the firemen reached the place and put down their long ladders they found him standing calmly in his wicker basket, entirely unhurt. The long, staunch keel, resting by its ends on the walls of the court, prevented him from being dashed to pieces. And so ended No. 5.

Most men would have given up aerial navigation after such an experience, but Santos-Dumont could not be deterred from continuing his experiments. The night of the very day which witnessed his fearful fall and the destruction of No. 5 he ordered a new balloon for "Santos-Dumont No. 6." It showed the pluck and determination of the man as nothing else could.

Twenty-two days after the aeronaut's narrow escape his new air-ship was finished and ready for a flight. No. 6 was practically the same as its predecessor—the triangular keel was retained, but an eighteen horse-power gasoline motor was substituted for the sixteen horse-power used previously. The propeller, made of silk stretched over a bamboo frame, was hung at the after end of the keel; the motor was a little aft of the centre, while the basket to which led the steering-gear, the emergency valve to the balloon, and the motor-controlling gear was suspended farther forward. To control the upward or downward pointing of the new air-ship, shifting ballast was used which ran along a wire under the keel from one end to the other; the cords controlling this ran to the basket also.

The new air-ship worked well, and the experimental flights were successful with one exception—when the balloon came in contact with a tree.

It was in October, 1901 (the 19th), when the Deutsch Prize Committee was asked to meet again and see a man try to drive a balloon against the wind, round the Eiffel Tower, and return.

The start took place at 2:42 P.M. of October 19, 1901, with a beam wind blowing. Straight as a bullet the air-ship sped for the steel shaft of the tower, rising as she flew. On and on she sped, while the spectators, remembering the finish of the last trial, watched almost breathlessly. With the air of a cup-racer turning the stake-boat she rounded the steel spire, a run of three and three-fifth miles, in nine minutes (at the rate of more than twenty-two miles an hour), and started on the home-stretch.

For a few moments all went well, then those who watched were horrified to see the propeller slow down and nearly stop, while the wind carried the air-ship toward the Tower. Just in time the motor was speeded up and the course was resumed. As the group of men watched the speck grow larger and larger until things began to take definite shape, the white blur of the whirling propeller could be seen and the small figure in the basket could be at last distinguished. Again the motor failed, the speed slackened, and the ship began to sink. Santos-Dumont threw out enough ballast to recover his equilibrium and adjusted the motor. With but three minutes left and some distance to go, the great dirigible balloon got up speed and rushed for the goal. At eleven and a half minutes past three, twenty-nine minutes and thirty-one seconds after starting, Santos-Dumont crossed the line, the winner of the Deutsch Prize. And so the young Brazilian accomplished that which had been declared impossible.

The following winter the aerial navigator, in the same No. 5, sailed many times over the waters of the Mediterranean from Monte Carlo. These flights over the water, against, athwart, and with the wind, some of them faster than the attending steamboats could travel, continued until through careless inflation of the balloon the air-ship and navigator sank into the sea. Santos-Dumont was rescued without being harmed in the least, and the air-ship was preserved intact, to be exhibited later to American sightseers.

"Santos-Dumont No. 6," the most successful of the series built by the determined Brazilian, looks as if it were altogether too frail to intrust with the carrying of a human being. The 105-foot-long balloon, a light yellow in colour, sways and undulates with every passing breeze. The steel piano wires by which the keel and apparatus are hung to the balloon skin are like spider-webs in lightness and delicacy, and the motor that has the strength of eighteen horses is hardly bigger than a barrel. A little forward of the motor is suspended to the keel the cigar-shaped gasoline reservoir, and strung along the top rod are the batteries which furnish the current to make the sparks for the purpose of exploding the gas in the motor.

Santos-Dumont himself says that the world is still a long way from practical, everyday aerial navigation, but he points out the apparent fact that the dirigible balloon in the hands of determined men will practically put a stop to war. Henri Rochefort has said: "The day when it is established that a man can direct an air-ship in a given direction and cause it to maneuver as he wills—there will remain little for the nations to do but to lay down their arms."

The man who has done so much toward the abolishing of war can rest well content with his work.

HOW A FAST TRAIN IS RUN

The conductor stood at the end of the train, watch in hand, and at the moment when the hands indicated the appointed hour he leisurely climbed aboard and pulled the whistle cord. A sharp, penetrating hiss of escaping air answered the pull, and the train moved out of the great train-shed in its race against time. It was all so easy and comfortable that the passengers never thought of the work and study that had been spent to produce the result. The train gathered speed and rushed on at an appalling rate, but the passengers did not realise how fast they were going unless they looked out of the windows and saw the houses and trees, telegraph poles, and signal towers flash by.

It is the purpose of this chapter to tell how high speed is attained without loss of comfort to the passengers—in other words, to tell how a fast train is run.

When the conductor pulled the cord at the rear end of the long train a whistling signal was thus given in the engine-cab, and the engineer, after glancing down the tracks to see that the signals indicated a clear track, pulled out the long handle of the throttle, and the great machine obeyed his will as a docile horse answers a touch on the rein. He opened the throttle-valve just a little, so that but little steam was admitted to the cylinders, and the pistons being pushed out slowly, the driving-wheels revolved slowly and the train started gradually. When the end of the piston stroke was reached the used steam was expelled into the smokestack, creating a draught which in turn strengthened the heat of the fire. With each revolution of the driving-wheels, each cylinder—there is one on each side of every locomotive—blew its steamy breath into the stack twice. This kept the fire glowing and made the chou-chou sound that everybody knows and every baby imitates.

As the train gathered speed the engineer pulled the throttle open wider and wider, the puffs in the short, stubby stack grew more and more frequent, and the rattle and roar of the iron horse increased.

Down in the pit of the engine-cab the fireman, a great shovel in his hands, stood ready to feed the ravenous fires. Every minute or two he pulled the chain and yanked the furnace door open to throw in the coal, shutting the door again after each shovelful, to keep the fire hot.

FIRING A FAST LOCOMOTIVE

The fireman on a fast locomotive is kept extremely busy, for he must keep the steam-pressure up to the required standard—150 or 200 pounds—no matter how fast the sucking cylinders may draw it out. He kept his eyes on the steam-gage most of the time, and the minute the quivering finger began to drop, showing reduced pressure, he opened the door to the glowing furnace and fed the fire. The steam-cylinders act on the boiler a good deal as a lung-tester acts on a human being; the cylinders draw out the steam from the boiler, requiring a roaring fire to make the vapour rapidly enough and keep up the pressure.

Though the engineer seemed to be taking it easily enough with his hand resting lightly on the reversing-lever, his body at rest, the fireman was kept on the jump. If he was not shovelling coal he was looking ahead for signals (for many roads require him to verify the engineer), or adjusting the valves that admitted steam to the train-pipes and heated the cars, or else, noticing that the water in the boiler was getting low—and this is one of his greatest responsibilities, which, however, the engineer sometimes shares—he turned on the steam in the injector, which forced the water against the pressure into the boiler. All these things he has to do repeatedly even on a short run.

The engineer—or "runner," as he is called by his fellows—has much to do also, and has infinitely greater responsibility. On him depends the safety and the comfort of the passengers to a large degree; he must nurse his engine to produce the greatest speed at the least cost of coal, and he must round the curves, climb the grades, and make the slow-downs and stops so gradually that the passengers will not be disturbed.

To the outsider who rides in a locomotive-cab for the first time it seems as if the engineer settles down to his real work with a sigh of relief when the limits of the city have been passed; for in the towns there are many signals to be watched, many crossings to be looked out for, and a multitude of moving trains, snorting engines, and tooting whistles to distract one's attention. The "runner," however, seemed not to mind it at all. He pulled on his cap a little more firmly, and, after glancing at his watch, reached out for the throttle handle. A very little pull satisfied him, and though the increase in speed was hardly perceptible, the more rapid exhaust told the story of faster movement. As the train sped on, the engineer moved the reversing-lever notch by notch nearer the centre of the guide. This adjusted the "link-motion" mechanism, which is operated by the driving-axle, and cut off the steam entering the cylinders in such a way that it expanded more fully and economically, thus saving fuel without loss of power.

When a station was reached, when a "caution" signal was displayed, or whenever any one of the hundred or more things occurred that might require a stop or a slow-down, the engineer closed down the throttle and very gradually opened the air-brake valve that admitted compressed air to the brake-cylinders, not only on the locomotive but on all the cars. The speed of the train slackened steadily but without jar, until the power of the compressed air clamped the brake-shoes on the wheels so tightly that they were practically locked and the train was stopped. By means of the air-brake the engineer had almost entire control of the train. The pump that compresses the air is on the engine, and keeps the pressure in the car and locomotive reservoirs automatically up to the required standard.

Each stage of every trip of a train not a freight is carefully charted, and the engineer is provided with a time-table that shows where his train should be at a given time. It is a matter of pride with the engineers of fast trains to keep close to their schedules, and their good records depend largely on this running-time, but delays of various kinds creep in, and in spite of their best efforts engineers are not always able to make all their schedules. To arrive at their destinations on time, therefore, certain sections must be covered in better than schedule time, and then great skill is required to get the speed without a sacrifice of comfort for the passenger.

To most travellers time is more valuable than money, and so everything about a train is planned to facilitate rapid travelling. Almost every part of a locomotive is controlled from the cab, which prevents unnecessary stopping to correct defects; from his seat the engineer can let the condensed water out of the cylinders; he can start a jet of steam in the stack and create a draft through the fire-box; by the pressure of a lever he is able to pour sand on a slippery track, or by the manipulation of another lever a snow-scraper is let down from the cowcatcher. The practised ear of a locomotive engineer often enables him to discover defects in the working of his powerful machine, and he is generally able, with the aid of various devices always on hand, to prevent an increase of trouble without leaving the cab.

As explained above, a fast run means the use of a great deal of steam and therefore water; indeed, the higher the speed the greater consumption of water. Often the schedules do not allow time enough to stop for water, and the consumption is so great that it is impossible to carry enough to keep the engine going to the end of the run. There are provided, therefore, at various places along the line, tanks eighteen inches to two feet wide, six inches deep, and a quarter of a mile long. These are filled with water and serve as long, narrow reservoirs, from which the locomotive-tenders are filled while going at almost full speed. Curved pipes are let down into the track-tank as the train speeds on, and scoop up the water so fast that the great reservoirs are very quickly filled. This operation, too, is controlled from the engine-cab, and it is one of the fireman's duties to let down the pipe when the water-signal alongside the track appears. The locomotive, when taking water from a track-tank, looks as if it was going through a river: the water is dashed into spray and flies out on either side like the waves before a fast boat. Trainmen tell the story of a tramp who stole a ride on the front or "dead" end platform of the baggage car of a fast train. This car was coupled to the rear end of the engine-tender; it was quite a long run, without stops, and the engine took water from a track-tank on the way. When the train stopped, the tramp was discovered prone on the platform of the baggage car, half-drowned from the water thrown back when the engine took its drink on the run.

"Here, get off!" growled the brakeman. "What are you doing there?"

"All right, boss," sputtered the tramp. "Say," he asked after a moment, "what was that river we went through a while ago?"

Though the engineer's work is not hard, the strain is great, and fast runs are divided up into sections so that no one engine or its runner has to work more than three or four hours at a time.

It is realised that in order to keep the trainmen—and especially the engineers—alert and keenly alive to their work and responsibilities, it is necessary to make the periods of labour short; the same thing is found to apply to the machines also—they need rest to keep them perfectly fit.

Before the engineer can run his train, the way must be cleared for him, and when the train starts out it becomes part of a vast system. Each part of this intricate system is affected by every other part, so each train must run according to schedule or disarrange the entire plan.

Each train has its right-of-way over certain other trains, and the fastest train has the right-of-way over all others. If, for any reason, the fastest train is late, all others that might be in the way must wait till the flyer has passed. When anything of this sort occurs the whole plan has to be changed, and all trains have to be run on a new schedule that must be made up on the moment.

The ideal train schedules, or those by which the systems are regularly governed, are charted out beforehand on a ruled sheet, as a ship's course is charted on a voyage, in the main office of the railroad. Each engineer and conductor is provided with a printed copy in the form of a table giving the time of departure and arrival at the different points. When the trains run on time it is all very simple, and the work of the despatcher, the man who keeps track of the trains, is easy. When, however, the system is disarranged by the failure of a train to keep to its schedule, the despatcher's work becomes most difficult. From long training the despatchers become perfectly familiar with every detail of the sections of road under their control, the position of every switch, each station, all curves, bridges, grades, and crossings. When a train is delayed and the system spoiled, it is the despatcher's duty to make up another one on the spot, and arrange by telegrams, which are repeated for fear of mistakes, for the holding of this train and the movement of others until the tangle is straightened out. This problem is particularly difficult when a road has but one track and trains moving in both directions have to run on the same pair of rails. It is on roads of this sort that most of the accidents occur. Almost if not quite all depends on the clear-headedness and quick-witted grasp of the despatchers and strict obedience to orders by the trainmen. To remove as much chance of error as possible, safety signalling methods have been devised to warn the engineer of danger ahead. Many modern railroads are divided into short sections or "blocks," each of which is presided over by a signal-tower. At the beginning of each block stand poles with projecting arms that are connected with the signal-tower by wires running over pulleys. There are generally two to each track in each block, and when both are slanting downward the engineer of the approaching locomotive knows that the block he is about to enter is clear and also that the rails of the section before that is clear as well. The lower arm, or "semaphore," stands for the second block, and if it is horizontal the engineer knows that he must proceed cautiously because the second section already has a train in it; if the upper arm is straight the "runner" knows that a train or obstruction of some sort makes it unsafe to enter the first block, and if he obeys the strict rules he must stay where he is until the arm is lowered At night, red, white, and green lights serve instead of the arms: white, safety; green, caution; and red, danger. Accidents have sometimes occurred because the engineers were colour-blind and red and green looked alike to them. Most roads nowadays test all their engineers for this defect in vision.

In spite of all precautions, it sometimes happens that the block-signals are not set properly, and to avoid danger of rear-end collisions, conductors and brakemen are instructed (when, for any reason, their train stops where it is not so scheduled) to go back with lanterns at night, or flags by day, and be ready to warn any following train. If for any reason a train is delayed and has to move ahead slowly, torpedoes are placed on the track which are exploded by the engine that comes after and warn its engineer to proceed cautiously.

All these things the engineer must bear in mind, and beside his jockey-like handling of his iron horse, he must watch for signals that flash by in an instant when he is going at full speed, and at the same time keep a sharp lookout ahead for obstructions on the track and for damaged roadbed.

The conductor has nothing to do with the mechanical running of the train, though he receives the orders and is, in a general way, responsible. The passengers are his special care, and it is his business to see that their getting on and off is in accordance with their tickets. He is responsible for their comfort also, and must be an animated information bureau, loaded with facts about every conceivable thing connected with travel. The brakemen are his assistants, and stay with him to the end of the division; the engineer and fireman, with their engine, are cut off at the end of their division also.

The fastest train of a road is the pride of all its employees; all the trainmen aspire to a place on the flyer. It never starts out on any run without the good wishes of the entire force, and it seldom puffs out of the train-shed and over the maze of rails in the yard without receiving the homage of those who happen to be within sight. It is impossible to tell of all the things that enter into the running of a fast train, but as it flashes across States, intersects cities, thunders past humble stations, and whistles imperiously at crossings, it attracts the attention of all. It is the spectacular thing that makes fame for the road, appears in large type in the newspapers, and makes havoc with the time-tables, while the steady-going passenger trains and labouring freights do the work and make the money.

THIRTY YEARS' ADVANCE IN LOCOMOTIVE BUILDING

HOW AUTOMOBILES WORK

Every boy and almost every man has longed to ride on a locomotive, and has dreamed of holding the throttle-lever and of feeling the great machine move under him in answer to his will. Many of us have protested vigorously that we wanted to become grimy, hard-working firemen for the sake of having to do with the "iron horse."

It is this joy of control that comes to the driver of an automobile which is one of the motor-car's chief attractions: it is the longing of the boy to run a locomotive reproduced in the grown-up.

The ponderous, snorting, thundering locomotive, towering high above its steel road, seems far removed from the swift, crouching, almost noiseless motor-car, and yet the relationship is very close. In fact, the automobile, which is but a locomotive that runs at will anywhere, is the father of the greater machine.

About the beginning of 1800, self-propelled vehicles steamed along the roads of Old England, carrying passengers safely, if not swiftly, and, strange to say, continued to run more or less successfully until prohibited by law from using the highways, because of their interference with the horse traffic. Therefore the locomotive and the railroads throve at the expense of the automobile, and the permanent iron-bound right of way of the railroads left the highways to the horse.

The old-time automobiles were cumbrous affairs, with clumsy boilers, and steam-engines that required one man's entire attention to keep them going. The concentrated fuels were not known in those days, and heat-economising appliances were not invented.

It was the invention by Gottlieb Daimler of the high-speed gasoline engine, in 1885, that really gave an impetus to the building of efficient automobiles of all powers. The success of his explosive gasoline engine, forerunner of all succeeding gasoline motor-car engines, was the incentive to inventors to perfect the steam-engine for use on self-propelled vehicles.

Unlike a locomotive, the automobile must be light, must be able to carry power or fuel enough to drive it a long distance, and yet must be almost automatic in its workings. All of these things the modern motor car accomplishes, but the struggle to make the machinery more efficient still continues.

The three kinds of power used to run automobiles are steam, electricity, and gasoline, taken in the order of application. The steam-engines in motor-cars are not very different from the engines used to run locomotives, factory machinery, or street-rollers, but they are much lighter and, of course, smaller—very much smaller in proportion to the power they produce. It will be seen how compact and efficient these little steam plants are when a ten-horse-power engine, boiler, water-tank, and gasoline reservoir holding enough to drive the machine one hundred miles, are stored in a carriage with a wheel-base of less than seven feet and a width of five feet, and still leave ample room for four passengers.

It is the use of gasoline for fuel that makes all this possible. Gasoline, being a very volatile liquid, turns into a highly inflammable gas when heated and mixed with the oxygen in the air. A tank holding from twenty to forty gallons of gasoline is connected, through an automatic regulator which controls the flow of oil, to a burner under the boiler. The burner allows the oil, which turns into gas on coming in contact with its hot surface, to escape through a multitude of small openings and mix with the air, which is supplied from beneath. The openings are so many and so close together that the whole surface is practically one solid sheet of very hot blue flame. In getting up steam a separate blaze or flame of alcohol or gasoline is made, which heats the steel or iron with which the fuel-oil comes in contact until it is sufficiently hot to turn the oil to gas, after which the burner works automatically. A hand air-pump or one automatically operated by the engine maintains sufficient air pressure in the fuel-tank to keep a constant flow.

Most steam automobile boilers are of the water-tube variety—that is, water to be turned into steam is carried through the flames in pipes, instead of the heat in pipes through the water, as in the ordinary flue boilers. Compactness, quick-heating, and strength are the characteristics of motor-car boilers. Some of the boilers are less than twenty inches high and of the same diameter, and yet are capable of generating seven and one-half horse-power at a high steam pressure (150 to 200 pounds). In these boilers the heat is made to play directly on a great many tubes, and a full head of steam is generated in a few minutes. As the steam pressure increases, a regulator that shuts off the supply of gasoline is operated automatically, and so the pressure is maintained.

The water from which the steam is made is also fed automatically into the boiler, when the engine is in motion, by a pump worked by the engine piston. A hand-pump is also supplied by which the driver can keep the proper amount when the machine is still or in case of a breakdown. A water-gauge in plain sight keeps the driver informed at all times as to the amount of water in the boiler. From the boiler the steam goes through the throttle-valve—the handle of which is by the driver's side—direct to the engine, and there expands, pushes the piston up and down, and by means of a crank on the axle does its work.

The engines of modern automobiles are marvels of compactness—so compact, indeed, that a seven-horse-power engine occupies much less space than an ordinary barrel. The steam, after being used, is admitted to a coil of pipes cooled by the breeze caused by the motion of the vehicle, and so condensed into water and returned to the tank. The engine is started, stopped, slowed, and sped by the cutting off or admission of the steam through the throttle-valve. It is reversed by means of the same mechanism used on locomotives—the link-motion and reversing-lever, by which the direction of the steam is reversed and the engine made to run the other way.

After doing its work the steam is made to circulate round the cylinder (or cylinders, if there are more than one), keeping it extra hot—"superheated"; and thereafter it is made to perform a like duty to the boiler-feed water, before it is allowed to escape.

All steam-propelled automobiles, from the light steam runabout to the clumsy steam roller, are worked practically as described. Some machines are worked by compound engines, which simply use the power of expansion still left in the steam in a second larger cylinder after it has worked the first, in which case every ounce of power is extracted from the vapour.

The automobile builders have a problem that troubles locomotive builders very little—that is, compensating the difference between the speeds of the two driving-wheels when turning corners. Just as the inside man of a military company takes short steps when turning and the outside man takes long ones, so the inside wheel of a vehicle turns slowly while the outside wheel revolves quickly when rounding a corner. As most automobiles are propelled by power applied to the rear axle, to which the wheels are fixed, it is manifest that unless some device were made to correct the fault one wheel would have to slide while the other revolved. This difficulty has been overcome by cutting the axle in two and placing between the ends a series of gears which permit the two wheels to revolve at different speeds and also apply the power to both alike. This device is called a compensating gear, and is worked out in various ways by the different builders.

The locomotive builder accomplishes the same thing by making his wheels larger on the outside, so that in turning the wide curves of the railroad the whole machine slides to the inside, bringing to bear the large diameter of the outer wheel and the small diameter of the inner, the wheels being fixed to a solid axle.

The steam machine can always be distinguished by the thin stream of white vapour that escapes from the rear or underneath while it is in motion and also, as a rule, when it is at rest.

The motor of a steam vehicle always stops when the machine is not moving, which is another distinguishing feature, as the gasoline motors run continually, or at least unless the car is left standing for a long time.

As the owners of different makes of bicycles formerly wrangled over the merits of their respective machines, so now motor-car owners discuss the value of the different powers—steam, gasoline, and electricity.

Though steam was the propelling force of the earliest automobiles, and the power best understood, it was the perfection of the gasoline motor that revived the interest in self-propelled vehicles and set the inventors to work.

A gasoline motor is somewhat like a gun—the explosion of the gas in the motor-cylinder pushes the piston (which may be likened to the projectile), and the power thus generated turns a crank and drives the wheels.

The gasoline motor is the lightest power-generator that has yet been discovered, and it is this characteristic that makes it particularly valuable to propel automobiles. Santos-Dumont's success in aerial navigation is due largely to the gasoline motor, which generated great power in proportion to its weight.

A gasoline motor works by a series of explosions, which make the noise that is now heard on every hand. From the gasoline tank, which is always of sufficient capacity for a good long run, a pipe is connected with a device called the carbureter. This is really a gas machine, for it turns the liquid oil into gas, this being done by turning it into fine spray and mixing it with pure air. The gasoline vapour thus formed is highly inflammable, and if lighted in a closed space will explode. It is the explosive power that is made to do the work, and it is a series of small gun-fires that make the gasoline motor-car go.

All this sounds simple enough, but a great many things must be considered that make the construction of a successful working motor a difficult problem.

In the first place, the carbureter, which turns the oil into gas, must work automatically, the proper amount of oil being fed into the machine and the exact proportion of air admitted for the successful mixture. Then the gas must be admitted to the cylinders in just the right quantity for the work to be done. This is usually regulated automatically, and can also be controlled directly by the driver. Since the explosion of gas in the cylinder drives the piston out only, and not, as in the case of the steam-engine, back and forward, some provision must be made to complete the cycle, to bring back the piston, exhaust the burned gas, and refill the cylinder with a new charge.

In the steam-engine the piston is forced backward and forward by the expansive power of the steam, the vapour being admitted alternately to the forward and rear ends of the cylinder. The piston of the gasoline engine, however, working by the force of exploded gas, produces power when moving in one direction only—the piston-head is pushed out by the force of the explosion, just as the plunger of a bicycle pump is sometimes forced out by the pressure of air behind it. The piston is connected with the engine-crank and revolves the shaft, which is in turn connected with the driving-wheels. The movement of the piston in the cylinder performs four functions: first, the downward stroke, the result of the explosion of gas, produces the power; second, the returning up-stroke pushes out the burned gas; third, the next down-stroke sucks in a fresh supply of gas, which (fourth) is compressed by the following-up movement and is ready for the next explosion. This is called a two-cycle motor, because two complete revolutions are necessary to accomplish all the operations. Many machines are fitted with heavy fly-wheels, the swift revolution of which carries the impetus of the power stroke through the other three operations.

To keep a practically continuous forward movement on the driving-shaft, many motors are made with four cylinders, the piston of each being connected with the crank-shaft at a different angle, and each cylinder doing a different part of the work; for example, while No. 1 cylinder is doing the work from the force of the explosion, No. 2 is compressing, No. 3 is getting a fresh supply of gas, and No. 4 is cleaning out waste gas. A four-cylinder motor is practically putting forth power continuously, since one of the four pistons is always at work.

While this takes long to describe, the motion is faster than the eye can follow, and the "phut, phut" noise of the exhaust sounds like the tattoo of a drum. Almost every gasoline motor vehicle carries its own electric plant, either a set of batteries or more commonly a little magneto dynamo, which is run by the shaft of the motor. Electricity is used to make the spark that explodes the gas at just the right moment in the cylinders. All this is automatic, though sometimes the driver has to resort to the persuasive qualities of a monkey-wrench and an oil-can.

The exploding gas creates great heat, and unless something is done to cool the cylinders they get so hot that the gas is ignited by the heat of the metal. Some motors are cooled by a stream of water which, flowing round the cylinders and through coils of pipe, is blown upon by the breeze made by the movement of the vehicle. Others are kept cool by a revolving fan geared to the driving-shaft, which blows on the cylinders; while still others—small motors used on motor bicycles, generally—have wide ridges or projections on the outside of the cylinders to catch the wind as the machine rushes along.

The inventors of the gasoline motor vehicles had many difficulties to overcome that did not trouble those who had to deal with steam. For instance, the gasoline motor cannot be started as easily as a steam-engine. It is necessary to make the driving-shaft revolve a few times by hand in order to start the cylinders working in their proper order. Therefore, the motor of a gasoline machine goes all the time, even when the vehicle is at rest. Friction clutches are used by which the driving-shaft and the axles can be connected or disconnected at the will of the driver, so that the vehicle can stand while the motor is running; friction clutches are used also to throw in gears of different sizes to increase or decrease the speed of the vehicle, as well as to drive backward.

AN AUTOMOBILE BUCKBOARD

The early gasoline automobiles sounded, when moving, like an artillery company coming full tilt down a badly paved street. The exhausted gas coughed resoundingly, the gears groaned and shrieked loudly when improperly lubricated, and the whole machine rattled like a runaway tin-peddler. Ingenious mufflers have subdued the sputtering exhaust, the gears are made to run in oil or are so carefully cut as to mesh perfectly, rubber tires deaden the pounding of the wheels, and carefully designed frames take up the jar.

Steam and gasoline vehicles can be used to travel long distances from the cities, for water can be had and gasoline bought almost anywhere; but electric automobiles, driven by the third of the three powers used for self-propelled vehicles, must keep within easy reach of the charging stations.

Just as the perfection of the gasoline motor spurred on the inventors to adapt the steam-engine for use in automobiles, so the inventors of the storage battery, which is the heart of an electric carriage, were stirred up to make electric propulsion practical.

The storage battery of an electric vehicle is practically a tank that holds electricity; the electrical energy of the dynamo is transformed into chemical energy in the batteries, which in turn is changed into electrical energy again and used to run the motors.

Electric automobiles are the most simple of all the self-propelled vehicles. The current stored in the batteries is simply turned off and on the motors, or the pressure reduced by means of resistance which obstructs the flow, and therefore the power, of the current. To reverse, it is only necessary to change the direction of the current's flow; and in order to stop, the connection between motor and battery is broken by a switch.

Electricity is the ideal power for automobiles. Being clean and easily controlled, it seems just the thing; but it is expensive, and sometimes hard to get. No satisfactory substitute has been found for it, however, in the larger cities, and it may be that creative or "primary" batteries both cheap and effective will be invented and will do away with the one objection to electricity for automobiles.

The astonishing things of to-day are the commonplaces of to-morrow, and so the achievements of automobile builders as here set down may be greatly surpassed by the time this appears in print.

The sensations of the locomotive engineer, who feels his great machine strain forward over the smooth steel rails, are as nothing to the almost numbing sensations of the automobile driver who covered space at the rate of eighty-eight miles an hour on the road between Paris and Madrid: he felt every inequality in the road, every grade along the way, and each curve, each shadow, was a menace that required the greatest nerve and skill. Locomotive driving at a hundred miles an hour is but mild exhilaration as compared to the feelings of the motor-car driver who travels at fifty miles an hour on the public highway.

Gigantic motor trucks carrying tons of freight twist in and out through crowded streets, controlled by one man more easily than a driver guides a spirited horse on a country road.

Frail motor bicycles dash round the platter-like curves of cycle tracks at railroad speed, and climb hills while the riders sit at ease with feet on coasters.

An electric motor-car wends the streets of New York every day with thirty-five or forty sightseers on its broad back, while a groom in whipcord blows an incongruous coaching-horn in the rear.

Motor plows, motor ambulances, motor stages, delivery wagons, street-cars without tracks, pleasure vehicles, and even baby carriages, are to be seen everywhere.

In 1845, motor vehicles were forbidden the streets for the sake of the horses; in 1903, the horses are being crowded off by the motor-cars. The motor is the more economical—it is the survival of the fittest.

AN AUTOMOBILE PLOW

THE FASTEST STEAMBOATS

In 1807, the first practical steamboat puffed slowly up the Hudson, while the people ranged along the banks gazed in wonder. Even the grim walls of the Palisades must have been surprised at the strange intruder. Robert Fulton's *Clermont* was the forerunner of the fleets upon fleets of power-driven craft that have stemmed the currents of a thousand streams and parted the waves of many seas.

The *Clermont* took several days to go from New York to Albany, and the trip was the wonder of that time.

During the summer of 1902 a long, slim, white craft, with a single brass smokestack and a low deck-house, went gliding up the Hudson with a kind of crouching motion that suggested a cat ready to spring. On her deck several men were standing behind the pilot-house with stop-watches in their hands. The little craft seemed alive under their feet and quivered with eagerness to be off. The passenger boats going in the same direction were passed in a twinkling, and the tugs and sailing vessels seemed to dwindle as houses and trees seem to shrink when viewed from the rear platform of a fast train.

Two posts, painted white and in line with each other—one almost at the river's edge, the other 150 feet back—marked the starting-line of a measured mile, and were eagerly watched by the men aboard the yacht. She sped toward the starting-line as a sprinter dashes for the tape; almost instantly the two posts were in line, the men with watches cried "Time!" and the race was on. Then began such a struggle with Father Time as was never before seen; the wind roared in the ears of the passengers and snatched their words away almost before their lips had formed them; the water, a foam-flecked streak, dashed away from the gleaming white sides as if in terror. As the wonderful craft sped on she seemed to settle down to her work as a good horse finds himself and gets into his stride. Faster and faster she went, while the speed of her going swept off the black flume of smoke from her stack and trailed it behind, a dense, low-lying shadow.

"Look!" shouted one of the men into another's ear, and raised his arm to point. "We're beating the train!"

Sure enough, a passenger train running along the river's edge, the wheels spinning round, the locomotive throwing out clouds of smoke, was dropping behind. The train was being beaten by the boat. Quivering, throbbing with the tremendous effort, she dashed on, the water climbing her sides and lashing to spume at her stern.

"Time!" shouted several together, as the second pair of posts came in line, marking the finish of the mile. The word was passed to the frantically struggling firemen and engineers below, while those on deck compared watches.

"One minute and thirty-two seconds," said one.

"Right," answered the others.

Then, as the wonderful yacht *Arrow* gradually slowed down, they tried to realise the speed and to accustom themselves to the fact that they had made the fastest mile on record on water.

And so the *Arrow*, moving at the rate of forty-six miles an hour, followed the course of her ancestress, the *Clermont*, when she made her first long trip almost a hundred years before.

The *Clermont* was the first practical steamboat, and the *Arrow* the fastest, and so both were record-breakers. While there are not many points of resemblance between the first and the fastest boat, one is clearly the outgrowth of the other, but so vastly improved is the modern craft that it is hard to even trace its ancestry. The little *Arrow* is a screw-driven vessel, and her reciprocating engines—that is, engines operated by the pulling and pushing power of the steam-driven pistons in cylinders—developed the power of 4,000 horses, equal to 32,000 men, when making her record-breaking run. All this enormous power was used to produce speed, there being practically no room left in the little 130-foot hull for anything but engines and boilers.

There is little difference, except in detail, between the *Arrow's* machinery and an ordinary propeller tugboat. Her hull is very light for its strength, and it was so built as to slip easily through the water. She has twin engines, each operating its own shaft and propeller. These are quadruple expansion. The steam, instead of being allowed to escape after doing its work in the first cylinder, is turned into a larger one and then successively into two more, so that all of its expansive power is used. After passing through the four cylinders, the steam is condensed into water again by turning it into pipes around which circulates the cool water in which the vessel floats. The steam thus condensed to water is heated and pumped into the boiler, to be turned into steam, so the water has to do its work many times. All this saves weight and, therefore, power, for the lighter a vessel is the more easily she can be driven. The boilers save weight also by producing steam at the enormous pressure of 400 pounds to the square inch. Steadily maintained pressure means power; the greater the pressure the more the power. It was the inventive skill of Charles D. Mosher, who has built many fast yachts, that enabled him to build engines and boilers of great power in proportion to their weight. It was the ability of the inventor to build boilers and engines of 4,000 horse-power compact and light enough to be carried in a vessel 130 feet long, of 12 feet 6 inches breadth, and 3 feet 6 inches depth, that made it possible for the *Arrow* to go a mile in one minute and thirty-two seconds. The speed of the wonderful little American boat, however, was not the result of any new invention, but was due to the perfection of old methods.

In England, about five years before the *Arrow's* achievement, a little torpedo-boat, scarcely bigger than a launch, set the whole world talking by travelling at the rate of thirty-nine and three-fourths miles an hour. The little craft seemed to disappear in the white smother of her wake, and those who watched the speed trial marvelled at the railroad speed she made. The *Turbina*—for that was the little record-breaker's name—was propelled by a new kind of engine, and her speed was all the more remarkable on that account. C.A. Parsons, the inventor of the engine, worked out the idea that inventors have been studying for a long time—since 1629, in fact—that is, the rotary principle, or the rolling movement without the up-and-down driving mechanism of the piston.

The *Turbina* was driven by a number of steam-turbines that worked a good deal like the water-turbines that use the power of Niagara. Just as a water-wheel is driven by the weight or force of the water striking the blades or paddles of the wheel, so the force of the many jets of steam striking against the little wings makes the wheels of the steam-turbines revolve. If you take a card that has been cut to a circular shape and cut the edges so that little wings will be made, then blow on this winged edge, the card will revolve with a buzz; the Parsons steam-turbine works in the same way. A shaft bearing a number of steel disks or wheels, each having many wings set at an angle like the blades of a propeller, is enclosed by a drumlike casing. The disks at one end of the shaft are smaller than those at the other; the steam enters at the small end in a circle of jets that blow against the wings and set them and the whole shaft whirling. After passing the first disk and its little vanes, the steam goes through the holes of an intervening fixed partition that deflects it so that it blows afresh on the second, and so on to the third and fourth, blowing upon a succession of wheels, each set larger than the preceding one. Each of Parsons's steam-turbine engines is a series of turbines put in a steel casing, so that they use every ounce of the expansive power of the steam.

It will be noticed that the little wind-turbine that you blow with your breath spins very rapidly; so, too, do the wheels spun by the steamy breath of the boilers, and Mr. Parsons found that the propeller fastened to the shaft of his engine revolved so fast that a vacuum was formed around the blades, and its work was not half done. So he lengthened his shaft and put three propellers on it, reducing the speed, and allowing all of the blades to catch the water strongly.

The *Turbina*, speeding like an express train, glided like a ghost over the water; the smoke poured from her stack and the cleft wave foamed at her prow, but there was little else to remind her inventor that 2,300 horse-power was being expended to drive her. There was no jar, no shock, no thumping of cylinders and pounding of rapidly revolving cranks; the motion of the engine was rotary, and the propeller shafts, spinning at 2,000 revolutions per minute, made no more vibration than a windmill whirling in the breeze.

To stop the *Turbina* was an easy matter; Mr. Parsons had only to turn off the steam. But to make the vessel go backward another set of turbines was necessary, built to run the other way, and working on the same shaft. To reverse the direction, the steam was shut off the engines which revolved from right to left and turned on those designed to run backward, or from left to right. One set of the turbines revolved the propellers so that they pushed, and the other set, turning them the other way, pulled the vessel backward—one set revolving in a vacuum and doing no work, while the other supplied the power.

The Parsons turbine-engines have been used to propel torpedo-boats, fast yachts, and vessels built to carry passengers across the English Channel, and recently it has been reported that two new transatlantic Cunarders are to be equipped with them.

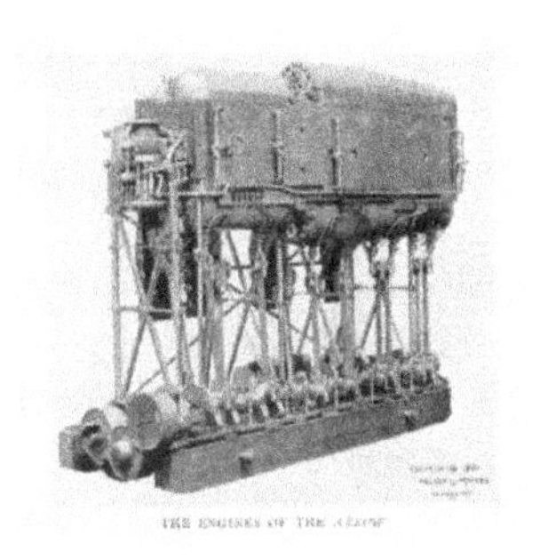

A few years after the Pilgrims sailed for the land of freedom in the tiny *Mayflower* a man named Branca built a steam-turbine that worked in a crude way on the same principle as Parsons's modern giant. The pictures of this first steam-turbine show the head and shoulders of a bronze man set over the flaming brands of a wood fire; his metallic lungs are evidently filled with water, for a jet of steam spurts from his mouth and blows against the paddles of a horizontal turbine wheel, which, revolving, sets in motion some crude machinery.

There is nothing picturesque about the steel-tube lungs of the boilers used by Parsons in the *Turbina* and the later boats built by him, and plain steel or copper pipes convey the steam to the whirling blades of the enclosed turbine wheels, but enormous power has been generated and marvellous speed gained. In the modern turbine a glowing coal fire, kept intensely hot by an artificial draft, has taken the place of the blazing sticks; the coils of steel tubes carrying the boiling water surrounded by flame replace the bronze-figure boiler, and the whirling, tightly jacketed turbine wheels, that use every ounce of pressure and save all the steam, to be condensed to water and used over again, have grown out of the crude machine invented by Branca.

As the engines of the *Arrow* are but perfected copies of the engine that drove the *Clermont*, so the power of the *Turbina* is derived from steam-motors that work on the same principle as the engine built by Branca in 1629, and his steam-turbine following the same old, old, ages old idea of the moss-covered, splashing, tireless water-wheel.

THE LIFE-SAVERS AND THEIR APPARATUS

Forming the outside boundary of Great South Bay, Long Island, a long row of sand-dunes faces the ocean. In summer groups of laughing bathers splash in the gentle surf at the foot of the low sand-hills, while the sun shines benignly over all. The irregular points of vessels' sails notch the horizon as they are swept along by the gentle summer breezes. Old Ocean is in a playful mood, and even children sport in his waters.

After the last summer visitor has gone, and the little craft that sail over the shallow bay have been hauled up high and dry, the pavilions deserted and the bathing-houses boarded up, the beaches take on a new aspect. The sun shines with a cold gleam, and the surf has an angry snarl to it as it surges up the sandy slopes and then recedes, dragging the pebbles after it with a rattling sound. The outer line of sand-bars, which in summer breaks the blue sea into sunny ripples and flashing whitecaps, then churns the water into fury and grips with a mighty hold the keel of any vessel that is unlucky enough to be driven on them. When the keen winter winds whip through the beach grasses on the dunes and throw spiteful handfuls of cutting sand and spray; when the great waves pound the beach and the crested tops are blown off into vapour, then the life-saver patrolling the beach must be most vigilant.

All along the coast, from Maine to Florida, along the Gulf of Mexico, the Great Lakes, and the Pacific, these men patrol the beach as a policeman walks his beat. When the winds blow hardest and sleet adds cutting force to the gale, then the surfmen, whose business it is to save life regardless of their own comfort or safety, are most alert.

All day the wind whistled through the grasses and moaned round the corners of the life-saving station; the gusts were cold, damp, and penetrating. With the setting of the sun there was a lull, but when the patrols started out at eight o'clock, on their four-hours' tour of duty, the wind had risen again and was blowing with renewed force. Separating at the station, one surf man went east and the other west, following the line of the surf-beaten beach, each carrying on his back a recording clock in a leather case, and also several candle-like Coston lights and a wooden handle.

A LIFE-SAVING CREW DRILLING WITH BEACH APPARATUS

"Wind's blowing some," said one of the men, raising his voice above the howl of the blast.

"Hope nothing hits the bar to-night," the other answered. Then both trudged off in opposite directions.

With pea-coats buttoned tightly and sou'westers tied down securely, the surfmen fought the gale on their watch-tour of duty. At the end of his beat each man stopped to take a key attached to a post, and, inserting it in the clock, record the time of his visit at that spot, for by this means is an actual record kept of the movements of the patrol at all times.

With head bent low in deference to the force of the blast, and eyes narrowed to slits, the surfman searched the seething sea for the shadowy outlines of a vessel in trouble.

Perchance as he looked his eye caught the dark bulk of a ship in a sea of foam, or the faint lines of spars and rigging through the spume and frozen haze—the unmistakable signs of a vessel in distress. An instant's concentrated gaze to make sure, then, taking a Coston signal from his pocket and fitting it to the handle, he struck the end on the sole of his boot. Like a parlour match it caught fire and flared out a brilliant red light. This served to warn the crew of the vessel of their danger, or notified them that their distress was observed and that help was soon forthcoming; it also served, if the surfman was near enough to the station, to notify the lookout there of the ship in distress. If the distance was too great or the weather too thick, the patrol raced back with all possible speed to the station and reported what he had seen. The patrol, through his long vigils under all kinds of weather conditions, learns every foot of his beat thoroughly, and is able to tell exactly how and where a stranded vessel lies, and whether she is likely to be forced over on to the beach or whether she will stick on the outer bar far beyond the reach of a line shot from shore.

In a few words spoken quickly and exactly to the point—for upon the accuracy of his report much depends—he tells the situation. For different conditions different apparatus is needed. The vessel reported one stormy winter's night struck on the shoal that runs parallel to the outer Long Island beach, far beyond the reach of a line from shore. Deep water lies on both sides of the bar, and after the shoal is passed the broken water settles down a little and gathers speed for its rush for the beach. These conditions were favourable for surf-boat work, and as the surfman told his tale the keeper or captain of the crew decided what to do.

The crew ran the ever-ready surf-boat through the double doors of its house down the inclined plane to the beach. Resting in a carriage provided with a pair of broad-tired wheels, the light craft was hauled by its sturdy crew through the clinging sand and into the very teeth of the storm to the point nearest the wreck.

The surf rolled in with a roar that shook the ground; fringed with foam that showed even through that dense midnight darkness, the waves were hungry for their prey. Each breaker curved high above the heads of the men, and, receding, the undertow sucked at their feet and tried to drag them under. It did not seem possible that a boat could be launched in such a sea. With scarcely a word of command, however, every man, knowing from long practice his position and specific duties, took his station on either side of the buoyant craft and, rushing into the surf, launched her; climbing aboard, every man took his appointed place, while the keeper, a long steering-oar in his hands, stood at the stern. All pulled steadily, while the steersman, with a sweep of his oar, kept her head to the seas and with consummate skill and judgment avoided the most dangerous crests, until the first watery rampart was passed. Adapting their stroke to the rough water, the six sturdy rowers propelled their twenty-five-foot unsinkable boat at good speed, though it seemed infinitely slow when they thought of the crew of the stranded vessel off in the darkness, helpless and hopeless. Each man wore a cork jacket, but in spite of their encumbrances they were marvellously active.

As is sometimes the case, before the surf-boat reached the distressed vessel she lurched over the bar and went driving for the beach.

The crew in the boat could do nothing, and the men aboard the ship were helpless. Climbing up into the rigging, the sailors waited for the vessel to strike the beach, and the life-savers put for shore again to get the apparatus needed for the new situation. To load the surf-boat with the wrecked, half-frozen crew of the stranded vessel, when there was none too much room for the oarsmen, and then encounter the fearful surf, was a method to be pursued only in case of dire need. To reach the wreck from shore was a much safer and surer method of saving life, not only for those on the vessel, but also for the surfmen.

The beach apparatus has received the greatest attention from inventors, since that part of the life-savers' outfit is depended upon to rescue the greatest number.

With a rush the surf-boat rolled in on a giant wave amid a smother of foam, and no sooner had her keel grated on the sand than her crew were out knee-deep in the swirling water and were dragging her up high and dry.

A minute later the entire crew, some pulling, some steering, dragged out the beach wagon. A light framework supported by two broad-tired wheels carried all the apparatus for rescue work from the beach. Each member of the crew had his appointed place and definite duties, according to printed instructions which each had learned by heart, and when the command was given every man jumped to his place as a well-trained man-of-war's-man takes his position at his gun.

Over hummocks of sand and wreckage, across little inlets made by the waves, in the face of blinding sleet and staggering wind, the life-savers dragged the beach wagon on the run.

Through the mist and shrouding white of the storm the outlines of the stranded vessel could just be distinguished.

Bringing the wagon to the nearest point, the crew unloaded their appliances.

Two men then unloaded a sand-anchor—an immense cross—and immediately set to work with shovels to dig a hole in the sand and bury it. While this was being done two others were busy placing a bronze cannon (two and one-half-inch bore) in position; another got out boxes containing small rope wound criss-cross fashion on wooden pins set upright in the bottom. The pins merely held the rope in its coils until ready for use, when board and pegs were removed. The free end of the line was attached to a ring in the end of the long projectile which the captain carried, together with a box of ammunition slung over his shoulders. The cylindrical projectile was fourteen and one-half inches long and weighed seventeen pounds. All these operations were carried on at once and with utmost speed in spite of the great difficulties and the darkness.

While the surf boomed and the wind roared, the captain sighted the gun—aided by Nos. 1 and 2 of the crew—aiming for the outstretched arms of the yards of the wrecked vessel. With the wind blowing at an almost hurricane rate, it was a difficult shot, but long practice under all kinds of difficulties had taught the captain just how to aim. As he pulled the lanyard, the little bronze cannon spit out fire viciously, and the long projectile, to which had been attached the end of the coiled line, sailed off on its errand of mercy. With a whir the line spun out of the box coil after coil, while the crew peered out over the breaking seas to see if the keeper's aim was true. At last the line stopped uncoiling and the life-savers knew that the shot had landed somewhere. For a time nothing happened, the slender rope reached out into the boiling waves, but no answering tugs conveyed messages to the waiting surfmen from the wrecked seamen.

At length the line began to slip through the fingers of the keeper who held it and moved seaward, so those on shore knew that the rope had been found and its use understood. The line carried out by the projectile served merely to drag out a heavy rope on which was run a sort of trolley carrying a breeches-buoy or sling.

The men on the wreck understood the use of the apparatus, or read the instructions printed in several languages with which the heavy rope was tagged. They made the end of the strong line fast to the mast well above the reach of the hungry seas, and the surfmen secured their end to the deeply buried sand-anchor, an inverted V-shaped crotch placed under the rope holding it above the water on the shore end. When this had been done, as much of the slack was taken up as possible, and the wreck was connected with the beach with a kind of suspension bridge.

All this occupied much time, for the hands of the sailors were numb with cold, the ropes stiff with ice, while the wild and angry wind snatched at the tackle and tore at the clinging figures.

In a trice the willing arms on shore hauled out the buoy by means of an endless line reaching out to the wreck and back to shore. Then with a joy that comes only to those who are saving a fellow-creature from death, the life-savers saw a man climb into the stout canvas breeches of the hanging buoy, and felt the tug on the whip-line that told them that the rescue had begun. With a will they pulled on the line, and the buoy, carrying its precious burden, rolled along the hawser, swinging in the wind, and now and then dipping the half-frozen man in the crests of the waves. It seemed a perilous journey, but as long as the wreck held together and the mast remained firmly upright the passengers on this improvised aerial railway were safe.

One after the other the crew were taken ashore in this way, the life-savers hauling the breeches-buoy forward and back, working like madmen to complete their work before the wreck should break up. None too soon the last man was landed, for he had hardly been dragged ashore when the sturdy mast, being able to stand the buffeting of the waves no longer, toppled over and floated ashore.

The life-savers' work is not over when the crew of a vessel is saved, for the apparatus must be packed on the beach wagon and returned to the station, while the shipwrecked crew is provided with dry clothing, fed, and cared for. The patrol continues on his beat throughout the night without regard to the hardships that have already been undergone.

The success of the surfmen in saving lives depends not only on their courage and strength, supplemented by continuous training which has been proved time and again, but the wonderful record of the life-saving service is due as well to the efficient appliances that make the work of the men effective.

Besides the apparatus already described, each station is provided with a kind of boat-car which has a capacity for six or seven persons, and is built so that its passengers are entirely enclosed, the hatch by which they enter being clamped down from the inside. When there are a great many people to be saved, this car is used in place of the breeches-buoy. It is hung on the hawser by rings at either end and pulled back and forth by the whip-line; or, if the masts of the vessel are carried away and there is nothing to which the heavy rope can be attached so that it will stretch clear above the wave-crests, in such an emergency the life-car floats directly on the water, and the whip-line is used to pull it to the shore with wrecked passengers and back to the wreck for more.

Everything that would help to save life under any condition is provided, and a number of appliances are duplicated in case one or more should be lost or damaged at a critical time. Signal flags are supplied, and the surfmen are taught their use as a means of communicating with people aboard a vessel in distress. Telephones connect the stations, so that in case of any special difficulty two or even three crews may be combined. When wireless telegraphy comes into general use aboard ship the stations will doubtless be equipped with this apparatus also, so that ships may be warned of danger.

The 10,000 miles of the United States ocean, gulf, and Great Lakes coasts, exclusive of Alaska and the island possessions, are guarded by 265 stations and houses of refuge at this writing, and new ones are added every year. Practically all of this immense coast-line is patrolled or watched over during eight or nine stormy months, and those that "go down to the sea in ships" may be sure of a helping hand in time of trouble.

The dangerous coasts are more thickly studded with stations, and the sections that are comparatively free from life-endangering reefs are provided with refuge houses where supplies are stored and where wrecked survivors may find shelter.

The Atlantic coast, being the most dangerous to shipping, is guarded by more than 175 stations; the Great Lakes require fifty or more to care for the survivors of the vessels that are yearly wrecked on their harbourless shores. For the Gulf of Mexico eight are considered sufficient, and the long Pacific coast also requires but eight.

The Life-Saving Service, formerly under the Treasury Department, now an important part of the Department of Commerce and Labour, was organised by Sumner I. Kimball, who was put at its head in 1871, and the great success and glory it has won is largely due to his energy and efficient enthusiasm.

The Life-Saving Service publishes a report of work accomplished through the year. It is a dry recital of facts and figures, but if the reader has a little imagination he can see the record of great deeds of heroism and self-sacrifice written between the lines.

As vessels labour through the wintry seas along our coasts, and the on-shore winds roar through the rigging, while the fog, mist or snow hangs like a curtain all around, it is surely a comfort to those at sea to know that all along the dangerous coast men specially trained, and equipped with the most efficient apparatus known, are always ready to stretch out a helping hand.

MOVING PICTURES

Some Strange Subjects and How They Were Taken

The grandstand of the Sheepshead Bay race-track, one spring afternoon, was packed solidly with people, and the broad, terra-cotta-coloured track was fenced in with a human wall near the judges' stand. The famous Suburban was to be run, and people flocked from every direction to see one of the greatest horse-races of the year. While the band played gaily, and the shrill cries of programme venders punctuated the hum of the voices of the multitude, and while the stable boys walked their aristocratic charges, shrouded in blankets, exercising them sedately—in the midst of all this movement, hubbub, and excitement a man a little to one side, apparently unconscious of all the uproar, was busy with a big box set up on a portable framework six or seven feet above the ground. The man was a new kind of photographer, and his big box was a camera with which he purposed to take a series of pictures of the race. Above the box, which was about two and a half feet square, was an electric motor from which ran a belt connecting with the inner mechanism; from the front of the box protruded the lens, its glassy eye so turned as to get a full sweep of the track; nearby on the ground were piled the storage batteries which were used to supply the current for the motor.

As the time for the race drew near the excitement increased, figures darted here, there and everywhere, the bobbing, brightly coloured hats of the women in the great slanting field of the grandstand suggesting bunches of flowers agitated by the breeze. Then the horses paraded in a thoroughbred fashion, as if they appreciated their lengthy pedigrees and understood their importance.

At last the splendid animals were lined up across the track, their small jockeys in their brilliantly coloured jackets hunched up like monkeys on their backs. Then the enormous crowd was quiet, the band was still, even the noisy programme venders ceased calling their wares, and the photographer stood quietly beside his camera, the motor humming, his hand on the switch that starts the internal machinery. Suddenly the starter dropped his arm, the barring gate flew up, and the horses sprang forward. "They're off!" came from a thousand throats in unison. The band struck up a lively air, and the vast assemblage watched with excited eyes the flying horses. As the horses swept on round the turn and down the back stretch the people seemed to be drawn from their seats, and by the time the racers made the turn leading into the home-stretch almost every one was standing and the roar of yelling voices was deafening.

All this time the photographer kept his eyes on his machine, which was rattling like a rapidly beaten drum, the cyclopean eye of the camera making impressions on a sensitised film-ribbon at the rate of forty a second, and every movement of the flying legs of the urging jockeys, even the puffs of dust that rose at the falling of each iron-shod hoof, was recorded for all time by the eye of the camera.

The horses entered the home-stretch and in a terrific burst of speed flashed by the throngs of yelling people and under the wire, a mere blur of shining bodies, brilliant colours of the jockeys' blouses, and yellow dust. The Suburban was over, and the great crowd that had come miles to see a race that lasted but a little more than two minutes (a grand struggle of giants, however), sank back into their seats or relaxed their straining gaze in a way that said plainer than words could say it, "It is over."

It was 4:45 in the afternoon. The photographer was all activity. The minute the race was over the motor above the great camera was stopped and the box was opened. From its dark interior another box about six inches square and two inches deep was taken: this box contained the record of the race, on a narrow strip of film two hundred and fifty feet long, the latent image of thousands of separate pictures.

Then began another race against time, for it was necessary to take that long ribbon across the city of Brooklyn, over the Bridge, across New York, over the North River by ferry to Hoboken on the Jersey side, develop, fix, and dry the two-hundred-and-fifty-foot-long film-negative, make a positive or reversed print on another two-hundred-and-fifty-foot film, carry it through the same photographic process, and show the spirited scene on the stereopticon screen of a metropolitan theatre the same evening.

That evening a great audience in the dark interior of a New York theatre sat watching a white sheet stretched across the stage; suddenly its white expanse grew dark, and against the background appeared "The Suburban, run this afternoon at 4:45 at Sheepshead Bay track; won by Alcedo, in 2 minutes 5 3-5 seconds."

Then appeared on the screen the picture of the scene that the thousands had travelled far to see that same afternoon. There were the wide, smooth track, the tower-like judges' stand, the oval turf of the inner field, and as the audience looked the starter moved his arm, and the rank of horses, life-size and quivering with excitement, shot forth. From beginning to end the great struggle was shown to the people seated comfortably in the city playhouse, several miles from the track where the race was run, just two hours and fifteen minutes after the winning horse dashed past the judges' stand. Every detail was reproduced; every movement of horses and jockeys, even the clouds of dust that rose from the hoof-beats, appeared clearly on the screen. And the audience rose gradually to their feet, straining forward to catch every movement, thrilled with excitement as were the mighty crowds at the actual race.

To produce the effect that made the people in the theatre forget their surroundings and feel as if they were actually overlooking the race-track itself, about five thousand separate photographs were shown.

It was discovered long ago that if a series of pictures, each of which showed a difference in the position of the legs of a man running, for instance, was passed quickly before the eye so that the space between the pictures would be screened, the figure would apparently move. The eyes retain the image they see for a fraction of a second, and if a new image carrying the movement a little farther along is presented in the same place, the eyes are deceived so that the object apparently actually moves. An ingenious toy called the zoltrope, which was based on this optical illusion, was made long before Edison invented the vitascope, Herman Caster the biograph and mutoscope, or the Lumiere brothers in France devised the cinematograph. All these different moving-picture machines work on the same principle, differing only in their mechanism.

A moving-picture machine is really a rapid-fire repeating camera provided with a lens allowing of a very quick exposure. Internal mechanism, operated by a hand-crank or electric motor, moves the unexposed film into position behind the lens and also opens and closes the shutter at just the proper moment. The same machinery feeds down a fresh section of the ribbon-like film into position and coils the exposed portion in a dark box, just as the film of a kodak is rolled off one spool and, after exposure, is wound up on another. The film used in the biograph when taking the Suburban was two and three-fourth inches wide and several hundred feet long; about forty exposures were made per second, and for each exposure the film had to come to a dead stop before the lens and then the shutter was opened, the light admitted for about one three-hundredth of a second, the shutter closed, and a new section of film moved into place, while the exposed portion was wound upon a spool in a light-tight box. The long, flexible film is perforated along both edges, and these perforations fit over toothed wheels which guide it down to the lens; the holes in the celluloid strip are also used by the feeding mechanism. In order that the interval between the pictures shall always be the same, the film must be held firmly in each position in turn; the perforations and toothed mechanism accomplish this perfectly.

In taking the picture of the Suburban race almost five thousand separate negatives (all on one strip of film, however) were made during the two minutes five and three-fifths seconds the race was being run. Each negative was perfectly clear, and each was different, though if one negative was compared to its neighbour scarcely any variance would be noted.

After the film has been exposed, the light-tight box containing it is taken out of the camera and taken to a gigantic dark-room, where it is wound on a great reel and developed, just as the image on a kodak film is brought out. The reel is hung by its axle over a great trough containing gallons of developer, so that the film wound upon it is submerged; and as the reel is revolved all of the sensitised surface is exposed to the action of the chemicals and gradually the latent pictures are developed. After the development has gone far enough, the reel, still carrying the film, is dipped in clean water and washed, and then a dip in a similar bath of clearing-and-fixing solution makes the negatives permanent—followed by a final washing in clean water. It is simply developing on a grand scale, thousands of separate pictures on hundreds of feet of film being developed at once.

A negative, however, is of no use unless a positive or print of some kind is made from it. If shown through a stereopticon, for instance, a negative would make all the shadows on the screen appear lights, and vice versa. A positive, therefore, is made by running a fresh film, with the negative, through a machine very much like the moving-picture camera. The unexposed surface is behind that of the negative, and at the proper intervals the shutter is opened and the admitted light prints the image of the negative on the unexposed film, just as a lantern slide is made, in fact, or a print on sensitised paper. The positives are made by this machine at the rate of a score or so in a second. Of course, the positive is developed in the same manner as the negative.

Therefore, in order to show the people in the theatre the Suburban, five hundred feet of film was exposed, developed, fixed, and dried, and nearly ten thousand separate and complete pictures were produced, in the space of two hours and fifteen minutes, including the time occupied in taking the films to and from the track, factory, and theatre.

Originally, successive pictures of moving objects were taken for scientific purposes. A French scientist who was studying aerial navigation set up a number of cameras and took successive pictures of a bird's flight. Doctor Muybridge, of Philadelphia, photographed trotting horses with a camera of his own invention that made exposures in rapid succession, in order to learn the different positions of the legs of animals while in rapid motion.

A Frenchman also—M. Mach—photographed a plant of rapid growth twice a day from exactly the same position for fifty consecutive days. When the pictures were thrown on the screen in rapid order the plant seemed to grow visibly.

The moving pictures provide a most attractive entertainment, and it was this feature of the idea, undoubtedly, that furnished the incentive to inventors. The public is always willing to pay well for a good amusement.

The makers of the moving-picture films have photographic studios suitably lighted and fitted with all the necessary stage accessories (scenery, properties, etc.) where the little comedies shown on the screens of the theatres are acted for the benefit of the rapid-fire camera and its operators, who are often the only spectators. One of these studios in the heart of the city of New York is so brilliantly lighted by electricity that pictures may be taken at full speed, thirty to forty-five per second, at any time of day or night. Another company has an open-air gallery large enough for whole troops of cavalry to maneuver before the camera, or where the various evolutions of a working fire department may be photographed.

Of course, when the pictures are taken in a studio or place prepared for the work the photographic part is easy—the camera man sets up his machine and turns the crank while the performers do the rest. But some extra-ordinary pictures have been taken when the photographer had to seek his scene and work his machine under trying and even dangerous circumstances.

During the Boer War in South Africa two operators for the Biograph Company took their bulky machine (it weighed about eighteen hundred pounds) to the very firing-line and took pictures of battles between the British and the Burghers when they were exposed to the fire of both armies. On one occasion, in fact, the operator who was turning the mechanism—he sat on a bicycle frame, the sprocket of which was connected by a chain with the interior machinery—during a battle, was knocked from his place by the concussion of a shell that exploded nearby; nevertheless, the film was saved, and the same man rode on horseback nearly seventy-five miles across country to the nearest railroad point so that the precious photographic record might be sent to London and shown to waiting audiences there.

Pictures were taken by the kinetoscope showing an ascent of Mount Blanc, the operator of the camera necessarily making the perilous journey also; different stages of the ascent were taken, some of them far above the clouds. For this series of pictures a film eight hundred feet long was required, and 12,800 odd exposures or negatives were made.

Successive pictures have been taken at intervals during an ocean voyage to show the life aboard ship, the swing of the great seas, and the rolling and pitching of the steamer. The heave and swing of the steamer and the mountainous waves have been so realistically shown on the screen in the theatre that some squeamish spectators have been made almost seasick. It might be comforting to those who were made unhappy by the sight of the heaving seas to know that the operator who took one series of sea pictures, when lashed with his machine in the lookout place on the foremast of the steamer, suffered terribly from seasickness, and would have been glad enough to set his foot on solid ground; nevertheless, he stuck to his post and completed the series.

It was a biograph operator that was engaged in taking pictures of a fire department rushing to a fire. Several pieces of apparatus had passed—an engine, hook-and-ladder company, and the chief; the operator, with his (then) bulky apparatus, large camera, storage batteries, etc., stood right in the centre of the street, facing the stream of engines, hose-wagons, and fire-patrol men. In order to show the contrast, an old-time hand-pump engine, dragged by a dozen men and boys, came along at full speed down the street, and behind and to one side of them followed a two-horse hose-wagon, going like mad. The men running with the old-time engine, not realising how narrow the space was and unaware of the plunging horses behind, passed the biograph man on one side on the dead run. The driver of the rapidly approaching team saw that there was no room for him to pass on the other side of the camera man, and his horses were going too fast to stop in the space that remained. He had but an instant to decide between the dozen men and their antiquated machine and the moving-picture outfit. He chose the latter, and, with a warning shout to the photographer, bore straight down on the camera, which continued to do its work faithfully, taking dozens of pictures a second, recording even the strained, anxious expression on the face of the driver. The pole of the hose-wagon struck the camera-box squarely and knocked it into fragments, and the wheels passed quickly over the pieces, the photographer meanwhile escaping somehow. By some lucky chance the box holding the coiled exposed film came through the wreck unscathed.

When that series was shown on the screen in a theatre the audience saw the engine and hook-and-ladder in turn come nearer and nearer and then rush by, then the line of running men with the old engine, and then—and their flesh crept when they saw it—a team of plunging horses coming straight toward them at frightful speed. The driver's face could be seen between the horses' heads, distorted with effort and fear. Straight on the horses came, their nostrils distended, their great muscles straining, their fore hoofs striking out almost, it seemed, in the faces of the people in the front row of seats. People shrank back, some women shrieked, and when the plunging horses seemed almost on them, at the very climax of excitement, the screen was darkened and the picture blotted out. The camera taking the pictures had continued to work to the very instant it was struck and hurled to destruction.

In addition to the stereopticon and its attendant mechanism, which is only suitable when the pictures are to be shown to an audience, a machine has been invented for the use of an individual or a small group of people. In the mutoscope the positives or prints are made on long strips of heavy bromide paper, instead of films, and are generally enlarged; the strip is cut up after development and mounted on a cylinder, so they radiate like the spokes of a wheel, and are set in the same consecutive order in which they were taken. The thousands of cards bearing the pictures at the outer ends are placed in a box, so that when the wheel of pictures is turned, by means of a crank attached to the axle, a projection holds each card in turn before the lens through which the observer looks. The projection in the top of the box acts like the thumb turning the pages of a book. Each of the pictures is presented in such rapid succession that the object appears to move, just as the scenes thrown on the screen by a lantern show action.

The mutoscope widens the use of motion-photography infinitely. The United States Government will use it to illustrate the workings of many of its departments at the World's Fair at St. Louis: the life aboard war-ships, the handling of big guns, army maneuvers, the life-saving service, post-office workings, and, in fact, many branches of the government service will be explained pictorially by this means.

Agents for manufacturers of large machinery will be able to show to prospective purchasers pictures of their machines in actual operation.

Living, moving portraits have been taken, and by means of a hand machine can be as easily examined as pictures through a stereoscope. It is quite within the bounds of possibility that circulating libraries of moving pictures will be established, and that every public school will have a projecting apparatus for the use of films, and a stereopticon or a mutoscope. In fact, a sort of circulating library already exists, films or mutoscope pictures being rented for a reasonable sum; and thus many of the most important of the world's happenings may be seen as they actually occurred.

Future generations will have histories illustrated with vivid motion pictures, as all the great events of the day, processions, celebrations, battles, great contests on sea and land are now recorded by the all-seeing eye of the motion-photographer's camera.

BRIDGE BUILDERS AND SOME OF THEIR ACHIEVEMENTS

In the old days when Rome was supreme a Caesar decreed that a bridge should be built to carry a military road across a valley, or ordered that great stone arches should be raised to conduct a stream of water to a city; and after great toil, and at the cost of the lives of unnumbered labourers, the work was done—so well done, in fact, that much of it is still standing, and some is still doing service.

In much the same regal way the managers of a railroad order a steel bridge flung across a chasm in the midst of a wilderness far from civilisation, or command that a new structure shall be substituted for an old one without disturbing traffic; and, lo and behold, it is done in a surprisingly short time. But the new bridges, in contrast to the old ones, are as spider webs compared to the overarching branches of a great tree. The old type, built of solid masonry, is massive, ponderous, while the new, slender, graceful, is built of steel.

One day a bridge-building company in Pennsylvania received the specifications giving the dimensions and particulars of a bridge that an English railway company wished to build in far-off Burma, above a great gorge more than eight hundred feet deep and about a half-mile wide. From the meagre description of the conditions and requirements, and from the measurements furnished by the railroad, the engineers of the American bridge company created a viaduct. Just as an author creates a story or a painter a picture, so these engineers built a bridge on paper, except that the work of the engineers' imagination had to be figured out mathematically, proved, and reproved. Not only was the soaring structure created out of bare facts and dry statistics, but the thickness of every bolt and the strain to be borne by every rod were predetermined accurately.

And when the plans of the great viaduct were completed the engineers knew the cost of every part, and felt so sure that the actual bridge in far-off Burma could be built for the estimated amount, that they put in a bid for the work that proved to be far below the price asked by English builders.

And so this company whose works are in Pennsylvania was awarded the contract for the Gokteik viaduct in Burma, half-way round the world from the factory.

BUILDING AN AMERICAN BRIDGE IN BURMAH

In the midst of a wilderness, among an ancient people whose language and habits were utterly strange to most Americans, in a tropical country where modern machinery and appliances were practically unknown, a small band of men from the young republic contracted to build the greatest viaduct the world had ever seen. All the material, all the tools and machinery, were to be carried to the opposite side of the earth and dumped on the edge of the chasm. From the heaps of metal the small band of American workmen and engineers, aided by the native labourers, were to build the actual structure, strong and enduring, that was conceived by the engineers and reduced to working-plans in far-off Pennsylvania.

From ore dug out of the Pennsylvania mountains the steel was made and, piece by piece, the parts were rolled, riveted, or welded together so that every section was exactly according to the measurements laid out on the plan. As each part was finished it was marked to correspond with the plan and also to show its relation to its neighbour. It was like a gigantic puzzle. The parts were made to fit each other accurately, so that when the workmen in Burma came to put them together the tangle of beams and rods, of trusses and braces should be assembled into a perfect, orderly structure—each part in its place and each doing its share of the work.

With men trained to work with ropes and tackle collected from an Indian seaport, and native riveters gathered from another place, Mr. J.C. Turk, the engineer in charge, set to work with the American bridgemen and the constructing engineer to build a bridge out of the pieces of steel that lay in heaps along the brink of the gorge. First, the traveller, or derrick, shipped from America in sections, was put together, and its long arm extended from the end of the tracks on which it ran over the abyss.

From above the great steel beams were lowered to the masonry foundations of the first tower and securely bolted to them, and so, piece by piece, the steel girders were suspended in space and swung this way and that until each was exactly in its proper position and then riveted permanently. The great valley resounded with the blows of hammers on red-hot metal, and the clangour of steel on steel broke the silence of the tropic wilderness. The towers rose up higher and higher, until the tops were level with the rim of the valley, and as they were completed the horizontal girders were built on them, the rails laid, and the traveller pushed forward until its arm swung over the foundation of the next tower.

And so over the deep valley the slender structure gradually won its way, supporting itself on its own web as it crawled along like a spider. Indeed, so tall were its towers and so slender its steel cords and beams that from below it appeared as fragile as a spider's web, and the men, poised on the end of swinging beams or standing on narrow platforms hundreds of feet in air, looked not unlike the flies caught in the web.

The towers, however, were designed to sustain a heavy train and locomotive and to withstand the terrific wind of the monsoon. The pressure of such a wind on a 320-foot tower is tremendous. The bridge was completed within the specified time and bore without flinching all the severe tests to which it was put. Heavy trains—much heavier than would ordinarily be run over the viaduct—steamed slowly across the great steel trestle while the railroad engineers examined with utmost care every section that would be likely to show weakness. But the designers had planned well, the steel-workers had done their full duty, and the American bridgemen had seen to it that every rivet was properly headed and every bolt screwed tight—and no fault could be found.

The bridge engineer's work is very diversified, since no two bridges are alike. At one time he might be ordered to span a stream in the midst of a populous country where every aid is at hand, and his next commission might be the building of a difficult bridge in a foreign wilderness far beyond the edge of civilisation.

Bridge-building is really divided into four parts, and each part requires a different kind of knowledge and experience.

First, the designer has to have the imagination to see the bridge as it will be when it is completed, and then he must be able to lay it out on paper section by section, estimating the size of the parts necessary for the stress they will have to bear, the weight of the load they will have to carry, the effect of the wind, the contraction and expansion of cold and heat, and vibration; all these things must be thought of and considered in planning every part and determining the size of each. Also he must know what kind of material to use that is best fitted to stand each strain, whether to use steel that is rigid or that which is so flexible that it can be tied in a knot. On the designer depends the price asked for the work, and so it is his business to invent, for each bridge is a separate problem in invention, a bridge that will carry the required weight with the least expenditure of material and labour and at the same time be strong enough to carry very much greater loads than it is ever likely to be called upon to sustain. The designer is often the constructor as well, and he is always a man of great practical experience. He has in his time stepped out on a foot-wide girder over a rushing stream, directing his men, and he has floundered in the mud of a river bottom in a caisson far below the surface of the stream, while the compressed air kept the ooze from flowing in and drowning him and his workmen.

The second operation of making the pieces that go into the structure is simply the following out of the clearly drawn plans furnished by the designing engineers. Different grades of steel and iron are moulded or forged into shape and riveted together, each part being made the exact size and shape required, even the position of the holes through which the bolts or rivets are to go that are to secure it to the neighbouring section being marked on the plan.

The foundations for bridges are not always put down by the builders of the bridge proper; that is a work by itself and requires special experience. On the strength and permanency of the foundation depends the life of the bridge. While the foundries and steel mills are making the metal-work the foundations are being laid. If the bridge is to cross a valley, or carry the roadway on the level across a depression, the placing of the foundations is a simple matter of digging or blasting out a big hole and laying courses of masonry; but if a pier is to be built in water, or the land on which the towers are to stand is unstable, then the problem is much more difficult.

For bridges like those that connect New York and Brooklyn, the towers of which rest on bed-rock below the river's bottom, caissons are sunk and the massive masonry is built upon them. If you take a glass and sink it in water, bottom up, carefully, so that the air will not escape, it will be noticed that the water enters the glass but a little way: the air prevents the water from filling the glass. The caisson works on the same principle, except that the air in the great boxlike chamber is highly compressed by powerful pumps and keeps the water and river ooze out altogether.

The caissons of the third bridge across the East River were as big as a good-sized house—about one hundred feet long and eighty feet wide. It took five large tugs more than two days to get one of them in its proper place. Anchored in its exact position, it was slowly sunk by building the masonry of the tower upon it, and when the lower edges of the great box rested on the bottom of the river men were sent down through an air-lock which worked a good deal like the lock of a canal. The men, two or three at a time, entered a small round chamber built of steel which was fitted with two air-tight doors at the top and bottom; when they were inside the air-lock, the upper door was closed and clamped tight, just as the gates leading from the lower level of a canal are closed after the boat is in the lock; then very gradually the air in the compartment is compressed by an air-compressor until the pressure in the air-lock is the same as that in the caisson chamber, when the lower door opened and allowed the men to enter the great dim room. Imagine a room eighty by one hundred feet, low and criss-crossed by massive timber braces, resting on the black, slimy mud of the river bottom; electric lights shine dimly, showing the half-naked workmen toiling with tremendous energy by reason of the extra quantity of oxygen in the compressed air. The workmen dug the earth and mud from under the iron-shod edges of the caisson, and the weight of the masonry being continually added to above sunk the great box lower and lower. From time to time the earth was mixed with water and sucked to the surface by a great pump. With hundreds of tons of masonry above, and the watery mud of the river on all sides far below the keels of the vessels that passed to and fro all about, the men worked under a pressure that was two or three times as great as the fifteen pounds to the square inch that every one is accustomed to above ground. If the pressure relaxed for a moment the lives of the men would be snuffed out instantly—drowned by the inrushing waters; if the excavation was not even all around, the balance of the top-heavy structure would be lost, the men killed, and the work destroyed entirely. But so carefully is this sort of work done that such an accident rarely occurs, and the caissons are sunk till they rest on bed-rock or permanent, solid ground, far below the scouring effect of currents and tides. Then the air-chamber is filled with concrete and left to support the great towers that pierce the sky above the waters.

THE SPIDER-WEB-LIKE VIADUCT ACROSS CAÑON DIABLO

The pneumatic tube, which is practically a steel caisson on a small scale operated in the same way, is often used for small towers, and many of the steel sky-scrapers of the cities are built on foundations of this sort when the ground is unstable.

Foundations of wooden and iron piles, driven deep in the ground below the river bottom, are perhaps the most common in use. The piles are sawed off below the surface of the water and a platform built upon them, which in turn serves as the foundation for the masonry.

The great Eads Bridge, which was built across the Mississippi at St. Louis, is supported by towers the foundations of which are sunk 107 feet below the ordinary level of the water; at this depth the men working in the caissons were subjected to a pressure of nearly fifty pounds to the square inch, almost equal to that used to run some steam-engines.

The bridge across the Hudson at Poughkeepsie was built on a crib or caisson open at the top and sunk by means of a dredge operated from above taking out the material from the inside. The wonder of this is hard to realise unless it is remembered that the steel hands of the dredge were worked entirely from above, and the steel rope sinews reached down below the surface more than one hundred feet sometimes; yet so cleverly was the work managed that the excavation was perfect all around, and the crib sank absolutely straight and square.

It is the fourth department of bridge-building that requires the greatest amount not only of knowledge but of resourcefulness. In the final process of erection conditions are likely to arise that were not considered when the plans were drawn.

The chief engineer in charge of the erection of a bridge far from civilisation is a little king, for it is necessary for him to have the power of an absolute monarch over his army of workmen, which is often composed of many different races.

With so many thousand tons of steel and stone dumped on the ground at the bridge site, with a small force of expert workmen and a greater number of unskilled labourers, in spite of bad weather, floods, or fearful heat, the constructing engineer is expected to finish the work within the specified time, and yet it must withstand the most exacting tests.

In the heart of Africa, five hundred miles from the coast and the source of supplies, an American engineer, aided by twenty-one American bridgemen, built twenty-seven viaducts from 128 to 888 feet long within a year.

The work was done in half the time and at half the cost demanded by the English bidders. Mr. Lueder, the chief engineer, tells, in his account of the work, of shooting lions from the car windows of the temporary railroad, and of seeing ostriches try to keep pace with the locomotive, but he said little of his difficulties with unskilled workmen, foreign customs, and almost unspeakable languages. The bridge engineer the world over is a man who accomplishes things, and who, furthermore, talks little of his achievements.

Though the work of the bridge builders within easy reach of the steel mills and large cities is less unusual, it is none the less adventurous.

In 1897, a steel arch bridge was completed that was built around the old suspension bridge spanning the Niagara River over the Whirlpool Rapids. The old suspension bridge had been in continuous service since 1855 and had outlived its usefulness. It was decided to build a new one on the same spot, and yet the traffic in the meantime must not be disturbed in the least. It would seem that this was impossible, but the engineers intrusted with the work undertook it with perfect confidence. To any one who has seen the rushing, roaring, foaming waters of unknown depth that race so fast from the spray-veiled falls that they are heaped up in the middle, the mere thought of men handling huge girders of steel above the torrent, and of standing on frail swinging platforms two hundred or more feet above the rapids, causes chills to run down the spine; yet the work was undertaken without the slightest doubt of its successful fulfilment.

It was manifestly impossible to support the new structure from below, and the old bridge was carrying about all it could stand, so it was necessary to build the new arch, without support from underneath, over the foaming water of the Niagara rapids two hundred feet below. Steel towers were built on either side of the gorge, and on them was laid the platform of the bridge from the towers nearest to the water around and under the old structure. The upper works were carried to the solid ground on a level with the rim of the gorge and there securely anchored with steel rods and chains held in masonry. Then from either side the arch was built plate by plate from above, the heavy sheets of steel being handled from a traveller or derrick that was pushed out farther and farther over the stream as fast as the upper platform was completed. The great mass of metal on both sides of the Niagara hung over the stream, and was only held from toppling over by the rods and chains solidly anchored on shore. Gradually the two ends of the uncompleted arch approached each other, the amount of work on each part being exactly equal, until but a small space was left between. The work was so carefully planned and exactly executed that the two completed halves of the arch did not meet, but when all was in readiness the chains on each side, bearing as they did the weight of more than 1,000,000 pounds, were lengthened just enough, and the two ends came together, clasping hands over the great gorge. Soon the tracks were laid, and the new bridge took up the work of the old, and then, piece by piece, the old suspension bridge, the first of its kind, was demolished and taken away.

Over the Niagara gorge also was built one of the first cantilever bridges ever constructed. To uphold it, two towers were built close to the water's edge on either side, and then from the towers to the shores, on a level with the upper plateau, the steel fabric, composed of slender rods and beams braced to stand the great weight it would have to carry, was built on false work and secured to solid anchorages on shore. Then on this, over tracks laid for the purpose, a crane was run (the same process being carried out on both sides of the river simultaneously), and so the span was built over the water 239 feet above the seething stream, the shore ends balancing the outer sections until the two arms met and were joined exactly in the middle. This bridge required but eight months to build, and was finished in 1883. From the car windows hardly any part of the slender structure can be seen, and the train seems to be held over the foaming torrent by some invisible support, yet hundreds of trains have passed over it, the winds of many storms have torn at its members, heat and cold have tried by expansion and contraction to rend it apart, yet the bridge is as strong as ever.

Sometimes bridges are built a span or section at a time and placed on great barges, raised to just their proper height, and floated down to the piers and there secured.

A railroad bridge across the Schuylkill at Philadelphia was judged inadequate for the work it had to do, and it was deemed necessary to replace it with a new one. The towers it rested upon, therefore, were widened, and another, stronger bridge was built alongside, the new one put upon rollers as was the old, and then between trains the old structure was pushed to one side, still resting on the widened piers, and the new bridge was pushed into its place, the whole operation occupying less than three minutes. The new replaced the old between the passing of trains that run at four or five-minute intervals. The Eads Bridge, which crosses the Mississippi at St. Louis, was built on a novel plan. Its deep foundations have already been mentioned. The great "Father of Waters" is notoriously fickle; its channel is continually changing, the current is swift, and the frequent floods fill up and scour out new channels constantly. It was necessary, therefore, in order to span the great stream, to place as few towers as possible and build entirely from above or from the towers themselves. It was a bold idea, and many predicted its failure, but Captain Eads, the great engineer, had the courage of his convictions and carried out his plans successfully. From each tower a steel arch was started on each side, built of steel tubes braced securely; the building on each side of every tower was carried on simultaneously, one side of every arch balancing the weight on the other side. Each section was like a gigantic seesaw, the tower acting as the centre support; the ends, of course, not swinging up and down. Gradually the two sections of every arch approached each other until they met over the turbid water and were permanently connected. With the completion of the three arches, built entirely from the piers supporting them, the great stream was spanned. The Eads Bridge was practically a double series of cantilevers balancing on the towers. Three arches were built, the longest being 520 feet long and the two shorter ones 502 feet each.

Every situation that confronts the bridge builder requires different handling; at one time he may be called upon to construct a bridge alongside of a narrow, rocky cleft over a rushing stream like the Royal Gorge, Colorado, where the track is hung from two great beams stretched across the chasm, or he may be required to design and construct a viaduct like that gossamer structure three hundred and five feet high and nearly a half-mile long across the Kinzua Creek, in Pennsylvania. Problems which have nothing to do with mechanics often try his courage and tax his resources, and many difficulties though apparently trivial, develop into serious troubles. The caste of the different native gangs who worked on the twenty-seven viaducts built in Central Africa is a case in point: each group belonging to the same caste had to be provided with its own quarters, cooking utensils, and camp furniture, and dire were the consequences of a mix-up during one of the frequent moves made by the whole party.

BEGINNING AN AMERICAN BRIDGE IN MID-AFRICA

ANOTHER VIEW OF THE GOKTEIK VIADUCT

And so the work of a bridge builder, whether it is creating out of a mere jumble of facts and figures a giant structure, the shaping of glowing metal to exact measurements, the delving in the slime under water for firm foundations, or the throwing of webs of steel across yawning chasms or over roaring streams, is never monotonous, is often adventurous, and in many, many instances is a great civilising influence.

SUBMARINES IN WAR AND PEACE

During the early part of the Spanish-American war a fleet of vessels patrolled the Atlantic coast from Florida to Maine. The Spanish Admiral Cervera had left the home waters with his fleet of cruisers and torpedo-boats and no one knew where they were. The lookouts on all the vessels were ordered to keep a sharp watch for strange ships, and especially for those having a warlike appearance. All the newspapers and letters received on board the different cruisers of the patrol fleet told of the anxiety felt in the coast towns and of the fear that the Spanish ships would appear suddenly and begin a bombardment. To add to the excitement and expectation, especially of the green crews, the men were frequently called out of their comfortable hammocks in the middle of the night, and sent to their stations at guns and ammunition magazines, just as if a battle was imminent; all this was for the purpose of familiarising the crews with their duties under war conditions, though no enlisted man knew whether he was called to quarters to fight or for drill.

These were the conditions, then, when one bright Sunday the crew of an auxiliary cruiser were very busy cleaning ship—a very thorough and absorbing business. While the men were in the thick of the scrubbing, one of the crew stood up to straighten his back, and looked out through an open port in the vessel's side. As he looked he caught a glimpse of a low, black craft, hardly five hundred yards off, coming straight for the cruiser. The water foamed at her bows and the black smoke poured out of her funnels, streaking behind her a long, sinister cloud. It was one of those venomous little torpedo-boats, and she was apparently rushing in at top speed to get within easy range of the large warship.

"A torpedo-boat is headed straight for us," cried the man at the port, and at the same moment came the call for general quarters.

As the men ran to their stations the word was passed from one to the other, "A Spanish torpedo-boat is headed for us."

With haste born of desperation the crew worked to get ready for action, and when all was ready, each man in his place, guns loaded, firing lanyards in hand, gun-trainers at the wheels, all was still—no command to fire was given.

From the signal-boys to the firemen in the stokehole—for news travels fast aboard ship—all were expecting the muffled report and the rending, tearing explosion of a torpedo under the ship's bottom. The terrible power of the torpedo was known to all, and the dread that filled the hearts of that waiting crew could not be put into words.

Of course it was a false alarm. The torpedo-boat flew the Stars and Stripes, but the heavy smoke concealed it, and the officers, perceiving the opportunities for testing the men, let it be believed that a boat belonging to the enemy was bearing down on them.

The crews of vessels engaged in future wars will have, not only swifter, surer torpedo-boats to menace them, but even more dreadful foes.

The conning towers of the submarines show but a foot or two above the surface—a sinister black spot on the water, like the dorsal fin of a shark, that suggests but does not reveal the cruel power below; for an instant the knob lingers above the surface while the steersman gets his bearings, and then it sinks in a swirling eddy, leaving no mark showing in what direction it has travelled. Then the crew of the exposed warship wait and wonder with a sickening cold fear in their hearts how soon the crash will come, and pray that the deadly submarine torpedo will miss its mark.

Submarine torpedo-boats are actual, practical working vessels to-day, and already they have to be considered in the naval plans for attack and defense.

Though the importance of submarines in warfare, and especially as a weapon of defense, is beginning to be thoroughly recognised, it took a long time to arouse the interest of naval men and the public generally sufficient to give the inventors the support they needed.

Americans once had within their grasp the means to blow some of their enemies' ships out of the water, but they did not realise it, as will be shown in the following, and for a hundred years the progress in this direction was hindered.

It was during the American Revolution that a man went below the surface of the waters of New York Harbour in a submarine boat just big enough to hold him, and in the darkness and gloom of the under-water world propelled his turtle-like craft toward the British ships anchored in mid-stream. On the outside shell of the craft rested a magazine with a heavy charge of gunpowder which the submarine navigator intended to screw fast to the bottom of a fifty-gun British man-of-war, and which was to be exploded by a time-fuse after he had got well out of harm's way.

Slowly and with infinite labour this first submarine navigator worked his way through the water in the first successful under-water boat, the crank-handle of the propelling screw in front of him, the helm at his side, and the crank-handle of the screw that raised or lowered the craft just above and in front. No other man had made a like voyage; he had little experience to guide him, and he lacked the confidence that a well-tried device assures; he was alone in a tiny vessel with but half an hour's supply of air, a great box of gunpowder over him, and a hostile fleet all around. It was a perilous position and he felt it. With his head in the little conning tower he was able to get a glimpse of the ship he was bent on destroying, as from time to time he raised his little craft to get his bearings. At last he reached his all-unsuspecting quarry and, sinking under the keel, tried to attach the torpedo. There in the darkness of the depths of North River this unnamed hero, in the first practical submarine boat, worked to make the first torpedo fast to the bottom of the enemy's ship, but a little iron plate or bolt holding the rudder in place made all the difference between a failure that few people ever heard of and a great achievement that would have made the inventor of the boat, David Bushnell, famous everywhere, and the navigator a great hero. The little iron plate, however, prevented the screw from taking hold, the tide carried the submarine past, and the chance was lost.

David Bushnell was too far ahead of his time, his invention was not appreciated, and the failure of his first attempt prevented him from getting the support he needed to demonstrate the usefulness of his under-water craft. The piece of iron in the keel of the British warship probably put back development of submarine boats many years, for Bushnell's boat contained many of the principles upon which the successful under-water craft of the present time are built.

One hundred and twenty-five years after the subsurface voyage described above, a steel boat, built like a whale but with a prow coming to a point, manned by a crew of six, travelling at an average rate of eight knots an hour, armed with five Whitehead torpedoes, and designed and built by Americans, passed directly over the spot where the first submarine boat attacked the British fleet.

The Holland boat *Fulton* had already travelled the length of Long Island Sound, diving at intervals, before reaching New York, and was on her way to the Delaware Capes.

She was the invention of John P. Holland, and the result of twenty-five years of experimenting, nine experimental boats having been built before this persistent and courageous inventor produced a craft that came up to his ideals. The cruise of the *Fulton* was like a march of triumph, and proved beyond a doubt that the Holland submarines were practical, sea-going craft.

At the eastern end of Long Island the captain and crew, six men in all, one by one entered the *Fulton* through the round hatch in the conning tower that projected about two feet above the back of the fish-like vessel. Each man had his own particular place aboard and definite duties to perform, so there was no need to move about much, nor was there much room left by the gasoline motor, the electric motor, storage batteries, air-compressor, and air ballast and gasoline tanks, and the Whitehead torpedoes. The captain stood up inside of the conning tower, with his eyes on a level with the little thick glass windows, and in front of him was the wheel connecting with the rudder that steered the craft right and left; almost at his feet was stationed the man who controlled the diving-rudders; farther aft was the engineer, all ready for the word to start his motor; another man controlled the ballast tanks, and another watched the electric motor and batteries.

With a clang the lid-like hatch to the conning tower was closed and clamped fast in its rubber setting, the gasoline engine began its rapid phut-phut, and the submarine boat began its long journey down Long Island Sound. The boat started in with her deck awash—that is, with two or three feet freeboard or of deck above the water-line. In this condition she could travel as long as her supply of gasoline held out—her tanks holding enough to drive her 560 knots at the speed of six knots an hour, when in the semi-awash condition; the lower she sank the greater the surface exposed to the friction of the water and the greater power expended to attain a given speed.

As the vessel jogged along, with a good part of her deck showing above the waves, her air ventilators were open and the burnt gas of the engine was exhausted right out into the open; the air was as pure as in the cabin of an ordinary ship. Besides the work of propelling the boat, the engine being geared to the electric motor made it revolve, so turning it into a dynamo that created electricity and filled up the storage batteries.

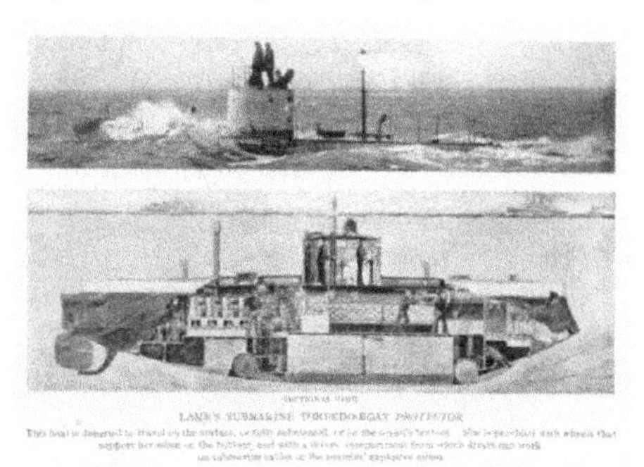

From time to time, as this whale-like ship plowed the waters of the Sound, a big wave would flow entirely over her, and the captain would be looking right into the foaming crest. The boat was built for under-water going, so little daylight penetrated the interior through the few small deadlights, or round, heavy glass windows, but electric incandescent bulbs fed by current from the storage batteries lit the interior brilliantly.

The boat had not proceeded far when the captain ordered the crew to prepare to dive, and immediately the engine was shut down and the clutch connecting its shaft with the electric apparatus thrown off and another connecting the electric motor with the propeller thrown in; a switch was then turned and the current from the storage batteries set the motor and propeller spinning. While this was being done another man was letting water into her ballast tanks to reduce her buoyancy. When all but the conning tower was submerged the captain looked at the compass to see how she was heading, noted that no vessels were near enough to make a submarine collision likely, and gave the word to the man at his feet to dive twenty feet. Then a strange thing happened. The diving-helmsman gave a twist to the wheel that connected with the horizontal rudders aft of the propeller, and immediately the boat slanted downward at an angle of ten degrees; the water rose about the conning tower until the little windows were level with the surface, and then they were covered, and the captain looked into solid water that was still turned yellowish-green by the light of the sun; then swiftly descending, he saw but the faintest gleam of green light coming through twenty feet of water. The *Fulton*, with six men in her, was speeding along at five knots an hour twenty feet below the shining waters of the Sound.

The diving-helmsman kept his eye on a gauge in front of him that measured the pressure of water at the varying depths, but the dial was so marked that it told him just how many feet the *Fulton* was below the surface. Another device showed whether the boat was on an even keel or, if not exactly, how many degrees she slanted up or down.

With twenty feet of salt water above her and as much below, this mechanical whale cruised along with her human freight as comfortable as they would have been in the same space ashore. The vessel contained sufficient air to last them several hours, and when it became vitiated there were always the tanks of compressed air ready to be drawn upon.

Except for the hum of the motor and the slight clank of the steering-gear, all was silent; none of the noises of the outer world penetrated the watery depths; neither the slap of the waves, the whir of the breeze, the hiss of steam, nor rattle of rigging accompanied the progress of this submarine craft. As silently as a fish, as far as the outer world was concerned, the *Fulton* crept through the submarine darkness. If an enemy's ship was near it would be an easy thing to discharge one of the five Whitehead torpedoes she carried and get out of harm's way before it struck the bottom of the ship and exploded.

In the tube which opened at the very tip end of the nose of the craft lay a Whitehead (or automobile) torpedo, which when properly set and ejected by compressed air propelled itself at a predetermined depth at a speed of thirty knots an hour until it struck the object it was aimed at or its compressed air power gave out.

The seven Holland boats built for the United States Navy, of which the *Fulton* is a prototype, carry five of these torpedoes, one in the tube and two on either side of the hold, and each boat is also provided with one compensating tank for each torpedo, so that when one or all are fired their weight may be compensated by filling the tanks with water so that the trim of the vessel will be kept the same and her stability retained.

The *Fulton*, however, was bent on a peaceful errand, and carried dummy torpedoes instead of the deadly engines of destruction that the man-o'-war's man dreads.

"Dive thirty," ordered the captain, at the same time giving his wheel a twist to direct the vessel's course according to the pointing finger of the compass.

"Dive thirty, sir," repeated the steersman below, and with a slight twist of his gear the horizontal rudders turned and the submarine inclined downward; the level-indicator showed a slight slant and the depth-gauge hand turned slowly round—twenty-two, twenty-five, twenty-eight, then thirty feet, when the helmsman turned his wheel back a little and the vessel forged ahead on a level keel.

At thirty feet below the surface the little craft, built like a cigar on purpose to stand a tremendous squeeze, was subjected to a pressure of 2,160 pounds to the square foot. To realise this pressure it will be necessary to think of a slab of iron a foot square and weighing 2,160 pounds pressing on every foot of the outer surface of the craft. Of course, the squeeze is exerted on all sides of the submarine boats when fully submerged, just as every one is subjected to an atmospheric pressure of fifteen pounds to the square inch on every inch of his body.

The *Fulton* and other submarine boats are so strongly built and thoroughly braced that they could stand an even greater pressure without damage.

When the commander of the *Fulton* ordered his vessel to the surface, the diving-steersman simply reversed his rudders so that they turned upward, and the propeller, aided by the natural buoyancy of the boat, simply pushed her to the surface. The Holland boats have a reserve buoyancy, so that if anything should happen to the machinery they would rise unaided to the surface.

Compressed air was turned into the ballast tanks, the water forced out so that the boat's buoyancy was increased, and she floated in a semi-awash, or light, condition. The engineer turned off the current from the storage batteries, threw off the motor from the propeller shaft, and connected the gasoline engine, started it up, and inside of five minutes from the time the *Fulton* was navigating the waters of the Sound at a depth of thirty feet she was sailing along on the surface like any other gasoline craft.

And so the ninety-mile journey down Long Island Sound, partly under water, partly on the surface, to New York, was completed. The greater voyage to the Delaware Capes followed, and at all times the little sixty-three-foot boat that was but eleven feet in diameter at her greatest girth carried her crew and equipment with perfect safety and without the least inconvenience.

Such a vessel, small in size but great in destructive power, is a force to be reckoned with by the most powerful battle-ship. No defense has yet been devised that will ward off the deadly sting of the submarine's torpedo, delivered as it is from beneath, out of the sight and hearing of the doomed ships' crews, and exploded against a portion of the hull that cannot be adequately protected by armour.

Though the conning-dome of a submarine presents a very small target, its appearance above water shows her position and gives warning of her approach. To avoid this tell-tale an instrument called a periscope has been invented, which looks like a bottle on the end of a tube; this has lenses and mirrors that reflect into the interior of the submarine whatever shows above water. The bottle part projects above, while the tube penetrates the interior.

The very unexpectedness of the submarine's attack, the mere knowledge that they are in the vicinity of a fleet and may launch their deadly missiles at any time, is enough to break down the nerves of the strongest and eventually throw into a panic the bravest crew.

That the crews of the war-ships will have to undergo the strain of submarine attack in the next naval war is almost sure. All the great nations of the world have built fleets of submarines or are preparing to do so.

In the development of under-water fighting-craft France leads, as she has the largest fleet and was the first to encourage the designing and building of them. But it was David Bushnell that invented and built the first practical working submarine boat, and in point of efficiency and practical working under service conditions in actual readiness for hostile action the American boats excel to-day.

A PEACEFUL SUBMARINE

Under the green sea, in the total darkness of the great depths and the yellowish-green of the shallows of the oceans, with the seaweeds waving their fronds about their barnacle-encrusted timbers and the creatures of the deep playing in and about the decks and rotted rigging, lie hundreds of wrecks. Many a splendid ship with a valuable cargo has gone down off a dangerous coast; many a hoard of gold or silver, gathered with infinite pains from the far corners of the earth, lies intact in decaying strong boxes on the bottom of the sea.

To recover the treasures of the deep, expeditions have been organised, ships have sailed, divers have descended, and crews have braved great dangers. Many great wrecking companies have been formed which accomplish wonders in the saving of wrecked vessels and cargoes. But in certain places all the time and at others part of the time, wreckers have had to leave valuable wrecks a prey to the merciless sea because the ocean is too angry and the waves too high to permit of the safe handling of the air-hose and life-line of the divers who are depended upon to do all the under-water work, rigging of hoisting-tackle, placing of buoys, etc. Indeed, it is often impossible for a vessel to stay in one place long enough to accomplish anything, or, in fact, to venture to the spot at all.

It was an American boy who, after reading Jules Verne's "Twenty Thousand Leagues Under the Sea," said to himself, "Why not?" and from that time set out to put into practice what the French writer had imagined.

Simon Lake set to work to invent a way by which a wrecked vessel or a precious cargo could be got at from below the surface. Though the waves may be tossing their whitecaps high in air and the strong wind may turn the watery plain into rolling hills of angry seas, the water twenty or thirty feet below hardly feels any surface motion. So he set to work to build a vessel that should be able to sail on the surface or travel on the bottom, and provide a shelter from which divers could go at will, undisturbed by the most tempestuous sea. People laughed at his idea, and so he found great difficulty in getting enough capital to carry out his plan, and his first boat, built largely with his own hands, had little in its appearance to inspire confidence in his scheme. Built of wood, fourteen feet long and five feet deep, fitted with three wheels, *Argonaut Junior* looked not unlike a large go-cart such as boys make out of a soap-box and a set of wooden wheels. The boat, however, made actual trips, navigated by its inventor, proving that his plan was feasible. *Argonaut Junior,* having served its purpose, was abandoned, and now lies neglected on one of the beaches of New York Bay.

The *Argonaut,* Mr. Lake's second vessel, had the regular submarine look, except that she was equipped with two great, rough tread-wheels forward, and to the underside of her rudder was pivoted another. She was really an under-water tricycle, a diving-bell, a wrecking-craft, and a surface gasoline-boat all rolled into one. When floating on the surface she looked not unlike an ordinary sailing craft; two long spars, each about thirty feet above the deck, forming the letter A—these were the pipes that admitted fresh air and discharged the burnt gases of the gasoline motor and the vitiated air that had been breathed. A low deck gave a ship-shape appearance when floating, but below she was shaped like a very fat cigar. Under the deck and outside of the hull proper were placed her gasoline tanks, safe from any possible danger of ignition from the interior. From her nose protruded a spar that looked like a bowsprit but which was in reality a derrick; below the derrick-boom were several glazed openings that resembled eyes and a mouth: these were the lookout windows for the under-water observer and the submarine searchlight.

The *Argonaut* was built to run on the surface or on the bottom; she was not designed to navigate half-way between. When in search of a wreck or made ready for a cruise along the bottom, the trap door or hatch in her turret-like pilot house was tightly closed; the water was let into her ballast tanks, and two heavy weights to which were attached strong cables that could be wound or unwound from the inside were lowered from their recesses in the fore and after part of the keel of the boat to the bottom; then the motor was started connected to the winding mechanism, and, the buoyancy of the boat being greatly reduced, she was drawn to the bottom by the winding of the anchor cables. As she sank, more and more water was taken into her tanks until she weighed slightly more than the water she displaced. When her wheels rested on the bottom her anchor-weights were pulled completely into their wells, so that they would not interfere with her movements.

Then the strange submarine vehicle began her voyage on the bottom of the bay or ocean. Since the pipes projected above the surface plenty of fresh air was admitted, and it was quite as easy to run the gasoline engine under water as on the surface. In the turrets, as far removed as possible from the magnetic influences of the steel hull, the compass was placed, and an ingeniously arranged mirror reflected its readings down below where the steersman could see it conveniently. Aft of the steering-wheel was the gasoline motor, connected with the propeller-shaft and also with the driving-wheels; it was so arranged that either could be thrown out of gear or both operated at once. She was equipped with depth-gauges showing the distance below the surface, and another device showing the trim of the vessel; compressed-air tanks, propelling and pumping machinery, an air-compressor and dynamo which supplied the current to light the ship and also for the searchlight which illuminated the under-water pathway—all this apparatus left but little room in the hold, but it was all so carefully planned that not an inch was wasted, and space was still left for her crew of three or four to work, eat, and even sleep, below the waves.

Forward of the main space of the boat were the diving and lookout compartments, which really were the most important parts of the boat, as far as her wrecking ability was concerned. By means of a trap door in the diving compartment through the bottom of the boat a man fitted with a diving-suit could go out and explore a wreck or examine the bottom almost as easily as a man goes out of his front door to call for an "extra." It will be thought at once, "But the water will rush in when the trap door is opened." This is prevented by filling the diving compartment, which is separated from the main part of the ship by steel walls, with compressed air of sufficient pressure to keep the water from coming in—that is, the pressure of water from without equals the pressure of air from within and neither element can pass into the other's domain.

An air-lock separates the diver's section from the main hold so that it is possible to pass from one to the other while the entrance to the sea is still open. A person entering the lock from the large room first closes the door between and then gradually admits the compressed air until the pressure is the same as in the diving compartment, when the door into it may be safely opened. When returning, this operation is simply reversed. The lookout stands forward of the diver's space. When the *Argonaut* rolls along the bottom, round openings protected with heavy glass permit the lookout to follow the beam of light thrown by the searchlight and see dimly any sizable obstruction. When the diving compartment is in use the man on lookout duty uses a portable telephone to tell his shipmates in the main room what is happening out in the wet, and by the same means the reports of the diver can be communicated without opening the air-lock.

This little ship (thirty-six feet long) has done wonderful things. She has cruised over the bottom of Chesapeake Bay, New York Bay, Hampton Roads, and the Atlantic Ocean, her driving-wheels propelling her when the bottom was hard, and her screw when the oozy condition of the submarine road made her spiked wheels useless except to steer with. Her passengers have been able to examine the bottom under twenty feet of water (without wetting their feet), through the trap door, with the aid of an electric light let down into the clear depths. Telephone messages have been sent from the bottom of Baltimore Harbour to the top of the New York *World* building, telling of the conditions there in contrast to the New York editor's aerial perch. Cables have been picked up and examined without dredging—a hook lowered through the trap door being all that was necessary. Wrecks have been examined and valuables recovered.

SINGING INTO THE TELEPHONE

Although the *Argonaut* travelled over 2,000 miles under water and on the surface, propelled by her own power, her inventor was not satisfied with her. He cut her in two, therefore, and added a section to her, making her sixty-six feet long; this allowed more comfortable quarters for her crew, space for larger engines, compressors, etc.

It was off Bridgeport, Connecticut, that the new *Argonaut* did her first practical wrecking. A barge loaded with coal had sunk in a gale and could not be located with the ordinary means. The *Argonaut*, however, with the aid of a device called the "wreck-detector," also invented by Mr. Lake, speedily found it, sank near it, and also submerged a new kind of freight-boat built for the purpose by the inventor. A diver quickly explored the hulk, opened the hatches of the freight-boat, which was cigar-shaped like the *Argonaut* and supplied with wheels so it could be drawn over the bottom, and placed the suction-tube in position. Seven minutes later eight tons of coal had been transferred from the wreck to the submarine freight-boat. The hatches were then closed and compressed air admitted, forcing out the water, and five minutes later the freight-boat was floating on the surface with eight tons of coal from a wreck which could not even be located by the ordinary means.

It is possible that in the future these modern "argonauts" will be seeking the golden fleeces of the sea in wrecks, in golden sands like the beaches of Nome, and that these amphibious boats will be ready along all the dangerous coasts to rush to the rescue of noble ships and wrest them from the clutches of the cruel sea.

Mr. Lake has also designed and built a submarine torpedo-boat that will travel on the surface, under the waves, or on the bottom; provided with both gasoline and electric power, and, fitted with torpedo discharge tubes, she will be able to throw a submarine torpedo; her diver could attach a charge of dynamite to the keel of an anchored warship, or she could do great damage by hooking up cables through her diver's trap door and cutting them, and by setting adrift anchored torpedoes and submarine mines.

Thus have Jules Verne's imaginings come true, and the dream *Nautilus,* whose adventures so many of us have breathlessly followed, has been succeeded by actual "Hollands" and practical "Argonauts" designed by American inventors and manned by American crews.

LONG-DISTANCE TELEPHONY

What Happens When You Talk into a Telephone Receiver

In Omaha, Nebraska, half-way across the continent and about forty hours from Boston by fast train, a man sits comfortably in his office chair and, with no more exertion than is required to lift a portable receiver off his desk, talks every day to his representative in the chief New England city. The man in Boston hears his chief's voice and can recognise the peculiarities in it just as if he stood in the same room with him. The man in Nebraska, speaking in an ordinary conversational tone, can be heard perfectly well in Boston, 1,400 miles away.

This is the longest talk on record—that is, it is the longest continuous telephone line in steady and constant use, though the human voice has been carried even greater distances with the aid of this wonderful instrument.

The telephone is so common that no one stops to consider the wonder of it, and not one person in a hundred can tell how it works.

At this time, when the telephone is as necessary as pen and ink, it is hard to realise a time when men could not speak to one another from a distance, yet a little more than a quarter of a century ago the genius who invented it first conceived the great idea.

Sometimes an inventor is a prophet: he sees in advance how his idea, perfected and in universal use, will change things, establish new manners and customs, new laws and new methods. Alexander Graham Bell was one of these prophetic inventors—the telephone was his invention, not his discovery. He first got the idea and then sought a way to make it practical. If you put yourself in his place, forget what has been accomplished, and put out of mind how the voice is transmitted from place to place by the slender wire, it would be impossible even then to realise how much in the dark Professor Bell was in 1874.

The human speaking voice is full of changes; unlike the notes from a musical instrument, there is no uniformity in it; the rise and fall of inflection, the varying sound of the vowels and consonants, the combinations of words and syllables—each produces a different vibration and different tone. To devise an instrument that would receive all these varying tones and inflections and change them into some other form of energy so that they could be passed over a wire, and then change them back to their original form, reproducing each sound and every peculiarity of the voice of the speaker in the ear of the hearer, was the task that Professor Bell set for himself. Just as you would sit down to add up a big column of figures, knowing that sooner or later you would get the correct answer, so he set himself to work out this problem in invention. The result of his study and determination is the telephones we use to-day. Many improvements have been invented by other men—Berliner, Edison, Blake, and others—but the idea and the working out of the principle is due to Professor Bell.

"CENTRAL" TELEPHONE OPERATORS AT WORK

Every telephone receiver and transmitter has a mouth-and ear-piece to receive or throw out the sound, a thin round sheet of lacquered metal—called a diaphragm, and an electromagnet; together they reproduce human speech. An electric current from a battery or from the central station flows continuously through the wires wound round the electromagnet in receiving and transmitting instruments, so when you speak into the black mouthpiece of the wall or desk receiver the vibrations strike against the thin sheet-iron diaphragm at the small end of the mouthpiece; the sound waves of the voice make it vibrate to a greater or less degree; the diaphragm is placed so that the core of the electromagnet is close to it, and as it vibrates the iron in it produces undulations (by induction) in the current which is flowing through the wires wound round the soft iron centre of the magnet. The wires of the coil are connected with the lines that go to the receiving telephone, so that this undulating current, coiling round the core of the magnet in the receiver, attracts and repels the iron of the diaphragm in it, and it vibrates just as the transmitter diaphragm did when spoken into; the undulating current is translated by it into words and sentences that have all the peculiarities of the original. And so when speaking into a telephone your voice is converted into undulations or waves in an electric current conveyed with incredible swiftness to the receiving instrument, and these are translated back into the vibrations that produce speech. This is really what takes place when you talk over a toy telephone made by a string stretched between the two tin mouth-pieces held at opposite sides of the room, with the difference that in the telephone the vibrations are carried electrically, while the toy carries them mechanically and not nearly so perfectly.

For once the world realised immediately the importance of a revolutionising invention, and telephone stations soon began to be established in the large cities. Quicker than the telegraph, for there was no need of an operator to translate the message, and more accurate, for if spoken clearly the words could be as clearly understood, the telephone service spread rapidly. Lines stretched farther and farther out from the central stations in the cities as improvements were invented, until the outlying wires of one town reached the outstretched lines of another, and then communication between town and town was established. Then two distant cities talked to each other through an intermediate town, and long-distance telephony was established. To-day special lines are built to carry long-distance messages from one great city to another, and these direct lines are used entirely except when storms break through or the rush of business makes the roundabout route through intermediate cities necessary.

As the nerves reaching from your finger-tips, from your ears, your eyes, and every portion of your body come to a focus in your brain and carry information to it about the things you taste, see, hear, feel, and smell, so the wires of a telephone system come together at the central station. And as it is necessary for your right hand to communicate with your left through your brain, so it is necessary for one telephone subscriber to connect through the central station with another subscriber.

The telephone has become a necessity of modern life, so that if through some means all the systems were destroyed business would be, for a time at least, paralysed. It is the perfection of the devices for connecting one subscriber with another, and for despatching the vast number of messages and calls at "central," that make modern telephony possible.

To handle the great number of spoken messages that are sent over the telephone wires of a great city it is necessary to divide the territory into districts, which vary in size according to the number of subscribers in them. Where the telephones are thickly installed the districts are smaller than in sections that are more sparsely settled.

Then all the telephone wires of a certain district converge at a central station, and each pair of wires is connected with its own particular switch at the switchboard of the station. That is simple enough; but when you come to consider that every subscriber must be so connected that he can be put into communication with every other subscriber, not only in his own section but also with every subscriber throughout the city, it will be seen that the switchboard at central is as marvellous as it is complicated. Some of the busy stations in New York have to take care of 6,000 or more subscribers and 10,000 telephone instruments, while the city proper is criss-crossed with more than 60,000 lines bearing messages from more than 100,000 "'phones." Just think of the babel entering the branch centrals that has to be straightened out and each separate series of voice undulations sent on its proper way, to be translated into speech again and poured into the proper ear. It is no wonder, then, that it has been found necessary to establish a school for telephone girls where they can be taught how to untangle the snarl and handle the vast, complicated system. In these schools the operators go through a regular course lasting a month. They listen to lectures and work out the instructions given them at a practice switchboard that is exactly like the service switchboard, except that the wires do not go outside of the building, but connect with the instructor's desk; the instructor calls up the pupils and sends messages in just the same way that the subscribers call "central" in the regular service.

At the terminal station of a great railroad, in the midst of a network of shining rails, stands the switchman's tower. By means of steel levers the man in his tower can throw his different switches and open one track to a train and close another; by means of various signals the switchman can tell if any given line is clear or if his levers do their work properly.

A telephone system may be likened, in a measure, to a complicated railroad line: the trunk wires to subscribers are like the tracks of the railroad, and the central station may be compared to the switch tower, while the central operators are like the switchmen. It is the central girls' business to see that connections are made quickly and correctly, that no lines are tied up unnecessarily, that messages are properly charged to the right persons, that in case of a break in a line the messages are switched round the trouble, and above all that there shall be no delay.

When you take your receiver off the hook a tiny electric bulb glows opposite the brass-lined hole that is marked with your number on the switchboard of your central, and the telephone girl knows that you are ready to send in a call—the flash of the little light is a signal to her that you want to be connected with some other subscriber. Whereupon, she inserts in your connection a brass plug to which a flexible wire is attached, and then opens a little lever which connects her with your circuit. Then she speaks into a kind of inverted horn which projects from a transmitter that hangs round her neck and asks: "Number, please?" You answer with the number, which she hears through the receiver strapped to her head and ear. After repeating the number the "hello" girl proceeds to make the connection. If the number required is in the same section of the city she simply reaches for the hole or connection which corresponds with it, with another brass plug, the twin of the one that is already inserted in your connection, and touches the brass lining with the plug. All the connections to each central station are so arranged and duplicated that they are within the reach of each operator. If the line is already "busy" a slight buzz is heard, not only by "central," but by the subscriber also if he listens; "central" notifies and then disconnects you. If the line is clear the twin plug is thrust into the opening, and at the same time "central" presses a button, which either rings a bell or causes a drop to fall in the private exchange station of the party you wish to talk to. The moment the new connection is made and the party you wish to talk to takes off the receiver from his hook, a second light glows beside yours, and continues to glow as long as the receiver remains off. The two little lamps are a signal to "central" that the connection is properly made and she can then attend to some other call. When your conversation is finished and your receivers are hung up the little lights go out. That signals "central" again, and she withdraws the plug from both holes and pushes another button, which connects with a meter made like a bicycle cyclometer. This little instrument records your call (a meter is provided for each subscriber) and at the same time lights the two tiny lamps again—a signal to the inspector, if one happens to be watching, that the call is properly recorded. All this takes long to read, but it is done in the twinkling of an eye. "Central's" hands are both free, and by long practice and close attention she is able to make and break connections with marvellous rapidity, it being quite an ordinary thing for an operator in a busy section to make ten connections a minute, while in an emergency this rate is greatly increased.

The call of one subscriber for another number in the same section, as described above—for instance, the call of 4341 Eighteenth Street for 2165 Eighteenth Street—is the easiest connection that "central" has to make.

As it is impossible for each branch exchange to be connected with every individual line in a great city, when a subscriber of one exchange wishes to talk with a subscriber of another, two central operators are required to make the connection. If No. 4341 Eighteenth Street wants to talk to 1748 Cortlandt Street, for instance, the Eighteenth Street central who gets the 4341 call makes a connection with the operator at Cortlandt Street and asks for No. 1748. The Cortlandt Street operator goes through the operation of testing to see if 1748 is busy, and if not she assigns a wire connecting the two exchanges, whereupon in Eighteenth Street one plug is put in 4341 switch hole; the twin plug is put into the switch hole connecting with the wire to Cortlandt Street; at Cortlandt Street the same thing is done with No. 1748 pair of plugs. The lights glow in both exchanges, notifying the operators when the conversation is begun and ended, and the operator of Eighteenth Street "central" makes the record in the same way as she does when both numbers are in her own district.

Besides the calls for numbers within the cities there are the out-of-town calls. In this case central simply makes connection with "Long Distance," which is a separate company, though allied with the city companies. "Long Distance" makes the connection in much the same way as the branch city exchanges. As the charges for long-distance calls depend on the length of the conversation, so the connection is made by an operator whose business it is to make a record of the length in minutes of the conversation and the place with which the city subscriber is connected. An automatic time stamp accomplishes this without possibility of error.

Sometimes the calls come from a pay station, in which case a record must be kept of the time occupied. This kind of call is indicated by the glow of a red light instead of a white one, and so "central" is warned to keep track, and the supervisors or monitors who constantly pass to and fro can note the kind of calls that come in, and so keep tab on the operators.

Other coloured lights indicate that the chief operator wishes to send out a general order and wishes all operators to listen. Another indicates that there is trouble somewhere on the line which needs the attention of the wire chief and repair department.

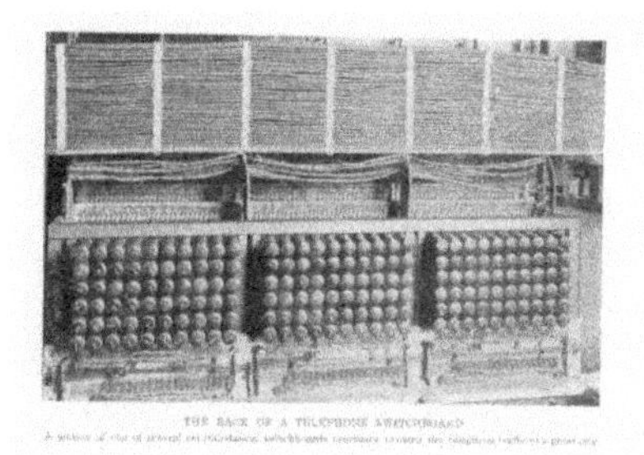

THE BACK OF A TELEPHONE SWITCHBOARD

The switchboards themselves are made of hard, black rubber, and are honeycombed with innumerable holes, each of which is connected with a subscriber. Below the switchboard is a broad shelf in which are set the miniature lamps and from which project the brass plugs in rows. The flexible cords containing the connecting wires are weighted and hang below, so that when a plug is pulled out of a socket and dropped it slides back automatically to its proper place, ready for use.

Many subscribers nowadays have their own private exchanges and several lines running to central. Perhaps No. 4341 Eighteenth Street, for instance, has 4342 and 4344 as well. This is indicated on the switchboard by a line of red or white drawn under the three switch-holes, so that central, finding one line busy, may be able to make connection with one of the other two, the line underneath showing at a glance which numbers belong to that particular subscriber.

If a subscriber is away temporarily, a plug of one colour is inserted in his socket, or if he is behind in his payments to the company a plug of another colour is put in, and if the service to his house is discontinued still another plug notifies the operator of the fact, and it remains there until that number is assigned to a new subscriber.

The operators sit before the switchboard in high swivel chairs in a long row, with their backs to the centre of the room.

From the rear it looks as if they were weaving some intricate fabric that unravels as fast as it is woven. Their hands move almost faster than the eye can follow, and the patterns made by the criss-crossed cords of the connecting plugs are constantly changing, varying from minute to minute as the colours in a kaleido-scope form new designs with every turn of the handle.

Into the exchange pour all the throbbing messages of a great city. Business propositions, political deals, scientific talks, and words of comfort to the troubled, cross and recross each other over the black switchboard. The wonder is that each message reaches the ear it was meant for, and that all complications, no matter how knotty, are immediately unravelled.

In the cities the telephone is a necessity. Business engagements are made and contracts consummated; brokers keep in touch with their associates on the floors of the exchanges; the patrolmen of the police force keep their chief informed of their movements and the state of the districts under their care; alarms of fire are telephoned to the fire-engine houses, and calls for ambulances bring the swift wagons on their errands of mercy; even wreckers telephone to their divers on the bottom of the bay, and undulating electrical messages travel to the tops of towering sky-scrapers.

A FEW TELEPHONE TRUNK WIRES
[illegible]

In Europe it is possible to hear the latest opera by paying a small fee and putting a receiver to your ear, and so also may lazy people and invalids hear the latest news without getting out of bed.

The farmers of the West and in eastern States, too, have learned to use the barbed wire that fences off their fields as a means of communicating with one another and with distant parts of their own property.

Mr. Pupin has invented an apparatus by which he hopes to greatly extend the distance over which men may talk, and it has even been suggested that Uncle Sam and John Bull may in the future swap stories over a transatlantic telephone line.

The marvels accomplished suggest the possible marvels to come. Automatic exchanges, whereby the central telephone operator is done away with, is one of the things that inventors are now at work on.

The one thing that prevents an unlimited use of the telephone is the expensive wires and the still more expensive work of putting them underground or stringing them overhead. So the capping of the climax of the wonders of the telephone would be wireless telephony, each instrument being so attuned that the undulations would respond only to the corresponding instrument. This is one of the problems that inventors are even now working upon, and it may be that wireless telephones will be in actual operation not many years after this appears in print.

A MACHINE THAT THINKS

A Typesetting Machine That Makes Mathematical Calculations

For many years it was thought impossible to find a short cut from author's manuscript to printing press—that is, to substitute a machine for the skilled hands that set the type from which a book or magazine is printed. Inventors have worked at this problem, and a number have solved it in various ways. To one who has seen the slow work of hand typesetting as the compositor builds up a long column of metal piece by piece, letter by letter, picking up each character from its allotted space in the case and placing it in its proper order and position, and then realises that much of the printed matter he sees is so produced, the wonder is how the enormous amount of it is ever accomplished.

In a page of this size there are more than a thousand separate pieces of type, which, if set by hand, would have to be taken one by one and placed in the compositor's "stick"; then when the line is nearly set it would have to be spaced out, or "justified," to fill out the line exactly. Then when the compositor's "stick" is full, or two and a half inches have been set, the type has to be taken out and placed in a long channel, or "galley." Each of these three operations requires considerable time and close application, and with each change there is the possibility of error. It is a long, expensive process.

A perfect typesetting machine should take the place of the hand compositor, setting the type letter by letter automatically in proper order at a maximum speed and with a minimum chance of error.

These three steps of hand composition, slow, expensive, open to many chances of mistake, have been covered at one stride at five times the speed, at one-third the cost, and much more accurately by a machine invented by Mr. Tolbert Lanston.

The operator of the Lanston machine sits at a keyboard, much like a typewriter in appearance, containing every character in common use (225 in all), and at a speed limited only by his dexterity he plays on the keys exactly as a typewriter works his machine. This is the sum total of human effort expended. The machine does all the rest of the work; makes the calculations and delivers the product in clean, shining new type, each piece perfect, each in its place, each line of exactly the right length, and each space between the words mathematically equal—absolutely "justified." It is practically hand composition with the human possibility of error, of weariness, of inattention, of ignorance, eliminated, and all accomplished with a celerity that is astonishing.

This machine is a type-casting machine as well as a typesetter. It casts the type (individual characters) it sets, perfect in face and body, capable of being used in hand composition or put to press directly from the machine and printed from.

As each piece of type is separate, alterations are easily made. The type for correction, which the machine itself casts for the purpose—a lot of a's, b's, etc.—is simply substituted for the words misspelled or incorrectly used, as in hand composition.

The Lanston machine is composed of two parts, the keyboard and the casting-setting machine. The keyboard part may be placed wherever convenient, away from noise or anything that is likely to distract or interrupt the operator, and the perforated roll of paper produced by it (which governs the setting machine) may be taken away as fast as it is finished. In the setting-casting machine is located the brains. The five-inch roll of paper, perforated by the keyboard machine (a hole for every letter), gives the signal by means of compressed air to the mechanism that puts the matrix (or type mould) in position and casts the type letter by letter, each character following the proper sequence as marked by the perforations of the paper ribbon. By means of an indicator scale on the keyboard the operator can tell how many spaces there are between the words of the line and the remaining space to be filled out to make the line the proper width. This information is marked by perforations on the paper ribbon by the pressure of two keys, and when the ribbon is transferred to the casting machine these space perforations so govern the casting that the line of type delivered at the "galley" complete shall be of exactly the proper length, and the spaces between the words be equal to the infinitesimal fraction of an inch.

The casting machine is an ingenious mechanism of many complicated parts. In a word, the melted metal (a composition of zinc and lead) is forced into a mold of the letter to be cast. Two hundred and twenty-five of these moulds are collected in a steel frame about three inches square, and cool water is kept circulating about them, so that almost immediately after the molten metal is injected into the lines and dots of the letter cut in the mould it hardens and drops into its slot, a perfect piece of type.

All this is accomplished at a rate of four or five thousand "ems" per hour of the size of type used on this page. The letter M is the unit of measurement when the amount of any piece of composition is to be estimated, and is written "em."

If this page were set by hand (taking a compositor of more than average speed as a basis for figuring), at least one hour of steady work would be required, but this page set by the Lanston machine (the operator being of the same grade as the hand compositor) would require hardly more than fifteen minutes from the time the manuscript was put into the operator's hands to the delivery complete of the newly cast type in galleys ready to be made up into pages, if the process were carried on continuously.

This marvellous machine is capable of setting almost any size of type, from the minute "agate" to and including "pica," a letter more than one-eighth of an inch high, and a line of almost any desired width, the change from one size to any other requiring but a few minutes. The Lanston machine sets up tables of figures, poetry, and all those difficult pieces of composition that so try the patience of the hand compositor.

It is called the monotype because it casts and sets up the type piece by piece.

Another machine, invented by Mergenthaler, practically sets up the moulds, by a sort of typewriter arrangement, for a line at a time, and then a casting is taken of a whole line at once. This machine is used much in newspaper offices, where the cleverness of the compositor has to be depended upon and there is little or no time for corrections. Several other machines set the regular type that is made in type foundries, the type being placed in long channels, all of the same sort, in the same grooves, and slipped or set in its proper place by the machine operated by a man at the keyboard. These machines require a separate mechanism that distributes each type in its proper place after use, or else a separate compositor must be employed to do this by hand. The machines that set foundry type, moreover, require a great stock of it, just as many hundred pounds of expensive type are needed for hand composition.

Though a machine has been invented that will put an author's words into type, no mechanism has yet been invented that will do away with type altogether. It is one of the problems still to be solved.

HOW HEAT PRODUCES COLD

ARTIFICIAL ICE-MAKING

One midsummers day a fleet of United States war-ships were lying at anchor in Guantanamo Bay, on the southern coast of Cuba. The sky was cloudless, and the tropic sun shone so fiercely on the decks that the bare-footed Jackies had to cross the unshaded spots on the jump to save their feet.

An hour before the quavering mess-call sounded for the midday meal, when the sun was shining almost perpendicularly, a boat's crew from one of the cruisers were sent over to the supply-ship for a load of beef. Not a breath was stirring, the smooth surface of the bay reflected the brazen sun like a mirror, and it seemed to the oarsmen that the salt water would scald them if they should touch it. Only a few hundred yards separated the two vessels, yet the heat seemed almost beyond endurance, and the shade cast by the tall steel sides of the supply-steamer, when the boat reached it, was as comforting as a cool drink to a thirsty man. The oars were shipped, and one man was left to fend off the boat while the others clambered up the swaying rope-ladder, crossed the scorching decks on the run, and went below. In two minutes they were in the hold of the refrigerator-ship, gathering the frost from the frigid cooling-pipes and snowballing each other, while the boat-keeper outside of the three-eighth-inch steel plating was fanning himself with his hat, almost dizzy from the quivering heat-waves that danced before his eyes. The great sides of beef, hung in rows, were frozen as hard as rock. Even after the strip of water had been crossed on the return journey and the meat exposed to the full, unobstructed glare of the sun the cruiser's messcooks had to saw off their portions, and the remainder continued hard as long as it lasted. But the satisfaction of the men who ate that fresh American beef cannot be told.

Cream from a famous dairy is sent to particular patrons in Paris, France, and it is known that in one instance, at least, a bottle of cream, having failed to reach the person to whom it was consigned, made the return transatlantic voyage and was received in New York three weeks after its first departure, perfectly sweet and good. Throughout the entire journey it was kept at freezing temperature by artificial means. These are but two striking examples of wonders that are performed every day.

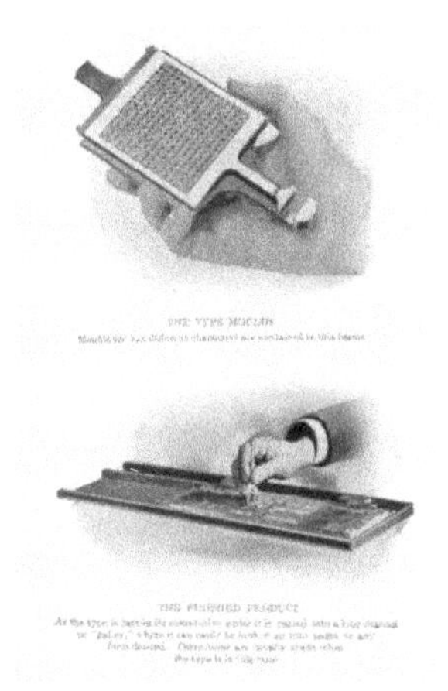

Cold, of course, is but the absence of heat, and so refrigerating machinery is designed to extract the heat from whatever substance it is desired to cool. The refrigerating agent used to extract the heat from the cold chamber must in turn have the heat extracted from it, and so the process must be continuous.

Water, when it boils and turns into steam or vapour, is heated by or extracts heat from the fire, but water vapourises at a high temperature and so cannot be used to produce cold. Other fluids are much more volatile and evaporate much more easily. Alcohol when spilt on the hand dries almost instantly and leaves a feeling of cold—the warmth of the hand boils the alcohol and turns it into vapour, and in doing so extracts the heat from the skin, making it cold; now, if the evaporated alcohol could be caught and compressed into its liquid form again you would have a refrigerating machine.

Alcohol is expensive and inflammable, and many other volatile substances have been discarded for the one or the other reason. Of all the fluids that have been tried, ammonia has been found to work most satisfactorily; it evaporates at a low temperature, is non-inflammable, and is comparatively cheap.

The hold of the supply-ship mentioned at the head of this chapter was a vast refrigerator, but no ice was used except that produced mechanically by the power in the ship. To produce the cold in the hold of the ship it was necessary to extract the heat in it; to accomplish this, coils ran round the space filled with cold brine, which, as it grew warm, drew the heat from the air. The brine in turn circulated through a tank containing pipes filled with ammonia vapour which extracted the heat from it; the brine then was ready to circulate through the coils in the hold again and extract more heat. The heat-extracting or cooling power of the ammonia is exerted continually by the process described below. Ammonia requires heat to expand and turn into vapour, and this heat it extracts from the substance surrounding it. In this marine refrigerating machine the ammonia got the heat from the brine in the tank, then it was drawn by a pump from the pipes in the tank, compressed by a power compressor, and forced into a second coil. The second coil is called a condenser because the vapour was there condensed into a fluid again. Over the pipes of the condenser cool water dripped constantly and carried off the heat in the ammonia vapour inside the coils and so condensed it into a fluid again—just as cold condenses steam into water. The compressor-pump then forced the fluid, ammonia through a small pipe from the condenser coils to the cooling coils in the tank of brine. The pipes of the cooling coils are much larger than those of the condenser, and as the fluid ammonia is forced in a fine spray into these large pipes and the pressure is relieved it expands or boils into the larger volume of vapour and in so doing extracts heat from the brine. The pump draws the heated vapour out, the compressor makes it dense, and the coils over which the cool water flows condenses it into fluid again, and so the circuit continues—through cooler, pump, compressor, and condenser, back into the cooling-tank.

In the meantime, the cold brine is being pumped through the pipes in the hold of the ship, where it extracts the heat from the air and the rows of sides of beef and then returns to the cooling-tank. In the refrigerating plant, then, of the supply-ship, there were two heat-extracting circuits, one of ammonia and the other of brine. Brine is used because it freezes at a very low temperature and continues to flow when unsalted water would be frozen solid. The ammonia is not used direct in the pipes in a big space like the hold of a ship, because so much of it would be required, and then there is always danger of the exposed pipes being broken and the dangerous fumes released.

Opposite as it may seem, heat is required to produce cold—for steam is necessary to drive the compressor and pump of a refrigerating plant, and fire of some sort is necessary to make steam.

The first artificial refrigerating machines produced cold by compressing and expanding air, the compressed air containing the heat being cooled by jets of cool water spirted into the cylinder containing it, then the compressed air was released or expanded into a larger chamber and thereby extracted the heat from brine or whatever substance surrounded it.

It is in the making of ice, however, that refrigerating machinery accomplishes its most surprising results. It was said in the writer's hearing recently that natural ice costs about as much when it was delivered at the docks or freight-yards of the large cities of the North as the product of the ice-machine. Of course, the manufactured ice is produced near the spot where it is consumed, and there is little loss through melting while it is being stored or transported, as in the case of the natural product.

There are two ways of making ice—or, rather, two methods using the same principle.

In the can system, a series of galvanized-iron cans about three and a half feet deep, eight inches wide, by two and a half feet long are suspended or rested in great tanks of brine connecting with the cooling-tank through which the pipes containing the ammonia vapour circulates. The vapour draws the heat from the brine, and the brine, which is kept moving constantly, in turn extracts the heat from the distilled water in the cans. While this method produces ice quickly, it is difficult to get ice of perfect clearness and purity, because the water in the can freezes on the sides, gradually getting thicker, retaining and concentrating in the centre any impurities that may be in the water. The finished cake, therefore, almost always has a white or cloudy appearance in the centre, and is frequently discolored.

In an ice-plant operated on the can system a great many blocks are freezing at once—in fact, the whole floor of a great room is honeycombed with trap-doors, a door for each can. The freezing is done in rotation, so that one group of cans is being emptied of their blocks of ice while others are still in process of congealing, while still others are being filled with fresh water. When the freezing is complete, jets of steam or quick immersion of the can in hot water releases the cake and the can is ready for another charge.

The plate system of artificial ice-making does away with the discoloration and the cloudiness, because the water containing the impurities or the air-bubbles is not frozen, but is drawn off and discarded.

In the plate system, great permanent tanks six feet deep and eight to twelve feet wide and of varying lengths are used. These tanks contain the clean, fresh water that is to be frozen into great slabs of ice. Into the tanks are sunk flat coils of pipe covered with smooth, metal plates on either side, and it is through these pipes that the ammonia vapour flows. The plates with the coils of pipe between them fit in the tank transversely, partitioning it off into narrow cells six feet deep, about twenty-two inches wide, and eight or ten feet long. In operation, the ammonia vapour flows through the pipes, chilling the plates and freezing the water so that a gradually thickening film of ice adheres to each side of each set of plates. As the ice gets thicker the unfrozen water between the slabs containing the impurities and air-bubbles gets narrower. When the ice on the plates is eight or ten inches thick very little of the unfrozen water remains between the great cakes, but it contains practically all the impurities. When the ice on the plates is thick enough, the ammonia vapour is turned off and steam forced through the pipes so the cakes come off readily, or else plates, cakes, and all are hoisted out of the tank and the ice melted off. The ice, clear and perfect, is then sawed into convenient sizes and shipped to consumers or stored for future use. Sometimes the plates or partitions are permanent, and, with the coils of pipes between them, cold brine is circulated, but in either case the two surfaces of ice do not come together, there being always a film of water between.

Still another method produces ice by forcing the clean water in extremely fine spray into a reservoir from which the air has been exhausted—into a vacuum, in other words; the spray condenses in the form of tiny particles of ice, which are attached to the walls of the reservoir. The ice grows thicker as a carpet of snow increases, one particle falling on and freezing to the others until the coating has reached the required thickness, when it is loosened and cut up in cakes of convenient size. The vacuum ice is of marble-like whiteness and appearance, but is perfectly pure, and it is said to be quite as hard.

More and more artificial ice is being used, even in localities where ice is formed naturally during parts of the year.

Many of the modern hotels are equipped with refrigerating plants where they make their own ice, cool their own storage-rooms, freeze the water in glass carafes for the use of their guests, and even cool the air that is circulated through the ventilating system in hot weather. In many large apartment-houses the refrigerators built in the various separate suites are kept at a freezing temperature by pipes leading to a refrigerating plant in the cellar. The convenience and neatness of this plan over the method of carrying dripping cakes from floor to floor in a dumb-waiter is evident.

Another use of refrigerating plants that is greatly appreciated is the making of artificial ice for skating-rinks. An artificial ice skating-rink is simply an ice machine on a grand scale—the ice being made in a great, thin, flat cake. Through the shallow tanks containing the fresh water coils of pipe through which flows the ammonia vapour or the cold brine are run from end to end or from side to side so that the whole bottom of the tank is gridironed with pipes, the water covering the pipes is speedily frozen, and a smooth surface formed. When the skaters cut up the surface it is flooded and frozen over again.

So efficient and common have refrigerating plants become that artificially cooled water is on tap in many public places in the great cities. Theatres are cooled during hot weather by a portion of the same machinery that supplies the heat in winter, and it is not improbable that every large establishment, private, or public, will in the near future have its own refrigerating plant.

Inventors are now at work on cold-air stoves that draw in warm air, extract the heat from it, and deliver it purified and cooled by many degrees.

Even the people of this generation, therefore, may expect to see their furnaces turned into cooling machines in summer. Then the ice-man will cease from troubling and the ice-cart be at rest.

Milton Keynes UK
Ingram Content Group UK Ltd.
UKHW030626061024
449204UK00004B/289